LECTURES ON
QUANTUM MECHANICS

Paul A. M. Dirac

DOVER PUBLICATIONS, INC.
Mineola, New York

Bibliographical Note

This Dover edition, first published in 2001, is an unabridged reprint of the work originally published by the Belfer Graduate School of Science, Yeshiva University, New York, in 1964.

Library of Congress Cataloging-in-Publication Data

Dirac, P. A. M. (Paul Adrien Maurice), 1902–
    Lectures on quantum mechanics / by Paul A.M. Dirac.
        p. cm.
    Originally published: New York : Belfer Graduate School of Science, Yeshiva University, 1964.
    ISBN 0-486-41713-1 (pbk.)
    1. Quantum theory. I. Title.

QC174.125 .D55 2001
530.12—dc21

00-065608

Manufactured in the United States of America
Dover Publications, Inc., 31 East 2nd Street, Mineola, N.Y. 11501

CONTENTS

Lecture No. 1

THE HAMILTONIAN METHOD

I'm very happy to be here at Yeshiva and to have this chance to talk to you about some mathematical methods that I have been working on for a number of years. I would like first to describe in a few words the general object of these methods.

In atomic theory we have to deal with various fields. There are some fields which are very familiar, like the electromagnetic and the gravitational fields; but in recent times we have a number of other fields also to concern ourselves with, because according to the general ideas of De Broglie and Schrödinger every particle is associated with waves and these waves may be considered as a field. So we have in atomic physics the general problem of setting up a theory of various fields in interaction with each other. We need a theory conforming to the principles of quantum mechanics, but it is quite a difficult matter to get such a theory.

One can get a much simpler theory if one goes over to the corresponding classical mechanics, which is the form which quantum mechanics takes when one makes Planck's constant $\hbar$ tend to zero. It is very much easier to visualize what one is doing in terms of classical mechanics. It will be mainly about classical mechanics that I shall be talking in these lectures.

Now you may think that that is really not good enough, because classical mechanics is not good enough to describe Nature. Nature is described by quantum mechanics. Why should one, therefore, bother so much about classical mechanics? Well, the quantum field theories are, as I said, quite difficult and so far, people have been able to build up quantum field theories only for fairly simple kinds of fields with simple interactions between them. It is quite possible that these simple fields with the simple interactions between them are not adequate for a description of Nature. The successes which we get with quantum field theories are rather limited. One is continually running into difficulties and one would like to broaden one's basis and have some possibility of bringing more general fields into account. For example, one would like to take into account the possibility that Maxwell's equations are not accurately valid. When one goes to distances very close to the charges that are producing the fields, one may have to modify Maxwell's field theory so as to make it into a non-linear electrodynamics. This is only one example of the kind of generalization which it is profitable to consider in our present state of ignorance of the basic ideas, the basic forces and the basic character of the fields of atomic theory.

In order to be able to start on this problem of dealing with more general fields, we must go over the classical theory. Now, if we can put the classical theory into the Hamiltonian form, then we can always apply certain standard rules so as to get a first approximation to a quantum theory. My talks will be mainly concerned with this problem of putting a general classical theory into the Hamiltonian form. When one has done that, one is well launched onto the path of getting an accurate quantum theory. One has, in any case, a first approximation.

Of course, this work is to be considered as a preliminary piece of work. The final conclusion of this piece of work must be to set up an accurate quantum theory, and that involves quite serious difficulties, difficulties of a fundamental character which people have been worrying over for quite a number of years. Some people are so much impressed by the difficulties of passing over from Hamiltonian classical mechanics to quantum mechanics that they think that maybe the whole method of working from Hamiltonian classical theory is a bad method. Particularly in the last few years people have been trying to set up alternative methods for getting quantum field theories. They have made quite considerable progress on these lines. They have obtained a number of conditions which have to be satisfied. Still I feel that these alternative methods, although they go quite a long way towards accounting for experimental results, will not lead to a final solution to the problem. I feel that there will always be something missing from them which we can only get by working from a Hamiltonian, or maybe from some generalization of the concept of a Hamiltonian. So I take the point of view that the Hamiltonian is really very important for quantum theory.

In fact, without using Hamiltonian methods one cannot solve some of the simplest problems in quantum theory, for example the problem of getting the Balmer formula for hydrogen, which was the very beginning of quantum mechanics. A Hamiltonian comes in therefore in very elementary ways and it seems to me that it is really quite essential to work from a Hamiltonian; so I want to talk to you about how far one can develop Hamiltonian methods.

I would like to begin in an elementary way and I take as my starting point an action principle. That is to say, I assume that there is an action integral which depends on the motion, such that, when one varies the motion, and puts down the conditions for the action integral to be stationary, one gets the equations of motion. The method of starting from an action principle has the one great advantage, that one can easily make the theory conform to the principle of relativity. We need our atomic theory to conform to relativity because in general we are dealing with particles moving with high velocities.

If we want to bring in the gravitational field, then we have to make our theory conform to the general principle of relativity, which means working with a space-time which is not flat. Now the gravitational field is not very important in atomic physics, because gravitational forces are extremely weak compared with the other kinds of forces which are present in atomic processes, and for practical purposes one can neglect the gravitational field. People have in recent years worked to some extent on bringing the gravitational field into the quantum theory, but I think that the main object of this work was the hope that bringing in the gravitational field might help to solve some of the difficulties. As far as one can see at present, that hope is not realized, and bringing in the gravitational field seems to add to the difficulties rather than remove them. So that there is not very much point at present in bringing gravitational fields into atomic theory. However, the methods which I am going to describe are powerful mathematical methods which would be available whether one brings in the gravitational field or not.

We start off with an action integral which I denote by

$$I = \int L \, dt. \qquad (1\text{-}1)$$

It is expressed as a time integral, the integrand L being the Lagrangian. So with an action principle we have a Lagrangian. We have to consider how to pass from that Lagrangian to a Hamiltonian. When we have got the Hamiltonian, we have made the first step toward getting a quantum theory.

You might wonder whether one could not take the Hamiltonian as the starting point and short-circuit this work of beginning with an action integral, getting a Lagrangian from it and passing from the Lagrangian to the Hamiltonian. The objection to trying to make this short-circuit is that it is not at all easy to formulate the conditions for a theory to be relativistic in terms of the Hamiltonian. In terms of the action integral, it is very easy to formulate the conditions for the theory to be relativistic: one simply has to require that the action integral shall be invariant. One can easily construct innumerable examples of action integrals which are invariant. They will automatically lead to equations of motion agreeing with relativity, and any developments from this action integral will therefore also be in agreement with relativity.

When we have the Hamiltonian, we can apply a standard method which gives us a first approximation to a quantum theory, and if we are lucky we might be able to go on and get an accurate quantum theory. You might again wonder whether one could not short-circuit that work to some extent. Could one not perhaps pass directly from the Lagrangian to the quantum theory, and short-circuit altogether the Hamiltonian ? Well, for some simple examples one *can* do that. For some of the simple fields which are used in physics the Lagrangian is quadratic in the velocities, and is like the Lagrangian which one has in the non-relativistic dynamics of particles. For these examples for which the Lagrangian is quadratic in the velocities, people have devised some methods for passing directly from the Lagrangian to the quantum theory. Still, this limitation of the Lagrangian's being quadratic in the velocities is quite a severe one. I want to avoid this limitation and to work with a Lagrangian which can be quite a general function of the velocities. To get a general formalism which will be applicable, for example, to the non-linear electrodynamics which I mentioned previously, I don't think one can in any way short-circuit the route of starting with an action integral, getting a Lagrangian, passing from the Langrangian to the Hamiltonian, and then passing from the Hamiltonian to the quantum theory. That is the route which I want to discuss in this course of lectures.

In order to express things in a simple way to begin with, I would like to start with a dynamical theory involving only a finite number of degrees of freedom, such as you are familiar with in particle dynamics. It is then merely a formal matter to pass from this finite number of degrees of freedom to the infinite number of degrees of freedom which we need for a field theory.

Starting with a finite number of degrees of freedom, we have dynamical coordinates which I denote by q.

The general one is q_n, $n = 1, \cdots, N$, N being the number of degrees of freedom. Then we have the velocities $dq_n/dt = \dot{q}_n$. The Lagrangian is a function $L = L(q, \dot{q})$ of the coordinates and the velocities.

You may be a little disturbed at this stage by the importance that the time variable plays in the formalism. We have a time variable t occurring already as soon as we introduce the Lagrangian. It occurs again in the velocities, and all the work of passing from Lagrangian to Hamiltonian involves one particular time variable. From the relativistic point of view we are thus singling out one particular observer and making our whole formalism refer to the time for this observer. That, of course, is not really very pleasant to a relativist, who would like to treat all observers on the same footing. However, it is a feature of the present formalism which I do not see how one can avoid if one wants to keep to the generality of allowing the Lagrangian to be *any* function of the coordinates and velocities. We can be sure that the contents of the theory are relativistic, even though the form of the equations is not manifestly relativistic on account of the appearance of one particular time in a dominant place in the theory.

Let us now develop this Lagrangian dynamics and pass over to Hamiltonian dynamics, following as closely as we can the ideas which one learns about as soon as one deals with dynamics from the point of view of working with general coordinates. We have the Lagrangian equations of motion which follow from the variation of the action integral:

$$\frac{d}{dt}\left(\frac{\partial L}{\partial \dot{q}_n}\right) = \frac{\partial L}{\partial q_n}. \tag{1-2}$$

[7]

To go over to the Hamiltonian formalism, we introduce the momentum variables p_n, which are defined by

$$p_n = \frac{\partial L}{\partial \dot{q}_n}. \tag{1-3}$$

Now in the usual dynamical theory, one makes the assumption that the momenta are independent functions of the velocities, but that assumption is too restrictive for the applications which we are going to make. We want to allow for the possibility of these momenta *not* being independent functions of the velocities. In that case, there exist certain relations connecting the momentum variables, of the type $\phi(q, p) = 0$.

There may be several independent relations of this type, and if there are, we distinguish them one from another by a suffix $m = 1, \cdots, M$, so we have

$$\phi_m(q, p) = 0. \tag{1-4}$$

The q's and the p's are the dynamical variables of the Hamiltonian theory. They are connected by the relations (1-4), which are called the *primary constraints* of the Hamiltonian formalism. This terminology is due to Bergmann, and I think it is a good one.

Let us now consider the quantity $p_n \dot{q}_n - L$. (Whenever there is a repeated suffix I assume a summation over all values of that suffix.) Let us make variations in the variables q and $\dot{q}$, in the coordinates and the velocities. These variations will cause variations to occur in the momentum variables p. As a result of these variations,

$$\delta(p_n \dot{q}_n - L)$$

$$= \delta p_n \dot{q}_n + p_n\, \delta \dot{q}_n - \left(\frac{\partial L}{\partial q_n}\right) \delta q_n - \left(\frac{\partial L}{\partial \dot{q}_n}\right) \delta \dot{q}_n$$

$$= \delta p_n \dot{q}_n - \left(\frac{\partial L}{\partial q_n}\right) \delta q_n \tag{1-5}$$

[8]

by (1-3). Now you see that the variation of this quantity $p_n\dot{q}_n - L$ involves only the variation of the q's and that of the p's. It does not involve the variation of the velocities. That means that $p_n\dot{q}_n - L$ can be expressed in terms of the q's and the p's, independent of the velocities. Expressed in this way, it is called the *Hamiltonian H*.

However, the Hamiltonian defined in this way is not uniquely determined, because we may add to it any linear combination of the ϕ's, which are zero. Thus, we could go over to another Hamiltonian

$$H^* = H + c_m\phi_m, \tag{1-6}$$

where the quantities c_m are coefficients which can be any function of the q's and the p's. H^* is then just as good as H; our theory cannot distinguish between H and H^*. The Hamiltonian is not uniquely determined.

We have seen in (1-5) that

$$\delta H = \dot{q}_n\,\delta p_n - \left(\frac{\partial L}{\partial q_n}\right)\delta q_n.$$

This equation holds for any variation of the q's and the p's subject to the condition that the constraints (1-4) are preserved. The q's and the p's cannot be varied independently because they are restricted by (1-4), but for any variation of the q's and the p's which preserves these conditions, we have this equation holding. From the general method of the calculus of variations applied to a variational equation with constraints of this kind, we deduce

$$\dot{q}_n = \frac{\partial H}{\partial p_n} + u_m\frac{\partial \phi_m}{\partial p_n} \tag{1-7}$$

and

$$-\frac{\partial L}{\partial q_n} = \frac{\partial H}{\partial q_n} + u_m\frac{\partial \phi_m}{\partial q_n}$$

[9]

or
$$\dot{p}_n = -\frac{\partial H}{\partial q_n} - u_m \frac{\partial \phi_m}{\partial q_n}, \qquad (1\text{-}8)$$

with the help of (1-2) and (1-3), where the u_m are unknown coefficients. We have here the *Hamiltonian equations of motion*, describing how the variables q and p vary in time, but these equations involve unknown coefficients u_m.

It is convenient to introduce a certain formalism which enables one to write these equations briefly, namely the Poisson bracket formalism. It consists of the following: If we have two functions of the q's and the p's, say $f(q, p)$ and $g(q, p)$, they have a *Poisson bracket* $[f, g]$ which is defined by

$$[f, g] = \frac{\partial f}{\partial q_n} \frac{\partial g}{\partial p_n} - \frac{\partial f}{\partial p_n} \frac{\partial g}{\partial q_n}. \qquad (1\text{-}9)$$

The Poisson brackets have certain properties which follow from their definition, namely $[f, g]$ is antisymmetric in f and g:

$$[f, g] = -[g, f]; \qquad (1\text{-}10)$$

it is linear in either member:

$$[f_1 + f_2, g] = [f_1, g] + [f_2, g], \text{ etc.}; \qquad (1\text{-}11)$$

and we have the product law,

$$[f_1 f_2, g] = f_1[f_2, g] + [f_1, g]f_2. \qquad (1\text{-}12)$$

Finally, there is the relationship, known as the *Jacobi Identity*, connecting three quantities:

$$[f, [g, h]] + [g, [h, f]] + [h, [f, g]] = 0. \qquad (1\text{-}13)$$

With the help of the Poisson bracket, one can rewrite

the equations of motion. For any function g of the q's and the p's, we have

$$\dot{g} = \frac{\partial g}{\partial q_n} \dot{q}_n + \frac{\partial g}{\partial p_n} \dot{p}_n. \tag{1-14}$$

If we substitute for q_n and p_n their values given by (1-7) and (1-8), we find that (1-14) is just

$$\dot{g} = [g, H] + u_m[g, \phi_m]. \tag{1-15}$$

The equations of motion are thus all written concisely in the Poisson bracket formalism.

We can write them in a still more concise formalism if we extend the notion of Poisson bracket somewhat. As I have defined Poisson brackets, they have a meaning only for quantities f and g which can be expressed in terms of the q's and the p's. Something more general, such as a general velocity variable which is not expressible in terms of the q's and p's, does not have a Poisson bracket with another quantity. Let us extend the meaning of Poisson brackets and suppose that they exist for any two quantities and that they satisfy the laws (1-10), (1-11), (1-12), and (1-13), but are otherwise undetermined when the quantities are not functions of the q's and p's.

Then we may write (1-15) as

$$\dot{g} = [g, H + u_m \phi_m]. \tag{1-16}$$

Here you see the coefficients u occurring in one of the members of a Poisson bracket. The coefficients u_m are *not* functions of the q's and the p's, so that we cannot use the definition (1-9) for determining the Poisson bracket in (1-16). However, we can proceed to work out

[11]

this Poisson bracket using the laws (1-10), (1-11), (1-12), and (1-13). Using the summation law (1-11) we have:

$$[g, H + u_m\phi_m] = [g, H] + [g, u_m\phi_m] \qquad (1\text{-}17)$$

and using the product law (1-12),

$$[g, u_m\phi_m] = [g, u_m]\phi_m + u_m[g, \phi_m]. \qquad (1\text{-}18)$$

The last bracket in (1-18) is well-defined, for g and ϕ_m are both functions of the q's and the p's. The Poisson bracket $[g, u_m]$ is *not* defined, but it is multiplied by something that vanishes, ϕ_m. So the first term on the right of (1-18) vanishes. The result is that

$$[g, H + u_m\phi_m] = [g, H] + u_m[g, \phi_m], \qquad (1\text{-}19)$$

making (1-16) agree with (1-15).

There is something that we have to be careful about in working with the Poisson bracket formalism: We have the constraints (1-4), but must not use one of these constraints *before* working out a Poisson bracket. If we did, we would get a wrong result. So we take it as a rule that Poisson brackets must all be worked out before we make use of the constraint equations. To remind us of this rule in the formalism, I write the constraints (1-4) as equations with a different equality sign $\approx$ from the usual. Thus they are written

$$\phi_m \approx 0. \qquad (1\text{-}20)$$

I call such equations weak equations, to distinguish them from the usual or strong equations.

One can make use of (1-20) only after one has worked out all the Poisson brackets which one is interested in. Subject to this rule, the Poisson bracket (1-19) is quite definite, and we have the possibility of writing our equations of motion (1-16) in a very concise form:

$$\dot{g} \approx [g, H_T] \qquad (1\text{-}21)$$

with a Hamiltonian I call the *total* Hamiltonian,

$$H_T = H + u_m \phi_m. \qquad (1\text{-}22)$$

Now let us examine the consequences of these equations of motion. In the first place, there will be some consistency conditions. We have the quantities ϕ which have to be zero throughout all time. We can apply the equation of motion (1-21) or (1-15) taking g to be one of the ϕ's. We know that $\dot{g}$ must be zero for consistency, and so we get some consistency conditions. Let us see what they are like. Putting $g = \phi_m$ and $\dot{g} = 0$ in (1-15), we have:

$$[\phi_m, H] + u_{m'}[\phi_m, \phi_{m'}] \approx 0. \qquad (1\text{-}23)$$

We have here a number of consistency conditions, one for each value of m. We must examine these conditions to see what they lead to. It is possible for them to lead directly to an inconsistency. They might lead to the inconsistency $1 = 0$. If that happens, it would mean that our original Lagrangian is such that the Lagrangian equations of motion are inconsistent. One can easily construct an example with just one degree of freedom. If we take $L = q$ then the Lagrangian equation of motion (1-2) gives immediately $1 = 0$. So you see, we cannot take the Lagrangian to be completely arbitrary. We must impose on it the condition that the Lagrangian equations of motion do not involve an inconsistency. With this restriction the equations (1-23) can be divided into three kinds.

[13]

One kind of equation reduces to $0 = 0$, i.e. it is identically satisfied, with the help of the primary constraints.

Another kind of equation reduces to an equation independent of the u's, thus involving only the q's and the p's. Such an equation must be independent of the primary constraints, otherwise it is of the first kind. Thus it is of the form

$$\chi(q, p) = 0. \qquad (1\text{-}24)$$

Finally, an equation in (1-23) may not reduce in either of these ways; it then imposes a condition on the u's.

The first kind we do not have to bother about any more. Each equation of the second kind means that we have another constraint on the Hamiltonian variables. Constraints which turn up in this way are called *secondary constraints*. They differ from the primary constraints in that the primary constraints are consequences merely of the equations (1-3) that define the momentum variables, while for the secondary constraints, one has to make use of the Lagrangian equations of motion as well.

If we have a secondary constraint turning up in our theory, then we get yet another consistency condition, because we can work out $\dot{\chi}$ according to the equation of motion (1-15) and we require that $\dot{\chi} \approx 0$. So we get another equation

$$[\chi, H] + u_m[\chi, \phi_m] \approx 0. \qquad (1\text{-}25)$$

This equation has to be treated on the same footing as (1-23). One must again see which of the three kinds it is. If it is of the second kind, then we have to push the process one stage further because we have a further secondary constraint. We carry on like that until we have exhausted all the consistency conditions, and the final result will be that we are left with a number of secondary constraints of the type (1-24) together with a number of conditions on the coefficients u of the type (1-23).

The secondary constraints will for many purposes be treated on the same footing as the primary constraints. It is convenient to use the notation for them:

$$\phi_k \approx 0, \quad k = M + 1, \ldots, M + K, \qquad (1\text{-}26)$$

where K is the total number of secondary constraints. They ought to be written as weak equations in the same way as primary constraints, as they are also equations which one must not make use of before one works out Poisson brackets. So all the constraints together may be written as

$$\phi_j \approx 0, \quad j = 1, \ldots, M + K \equiv \jmath. \qquad (1\text{-}27)$$

Let us now go over to the remaining equations of the third kind. We have to see what conditions they impose on the coefficients u. These equations are

$$[\phi_j, H] + u_m[\phi_j, \phi_m] \approx 0 \qquad (1\text{-}28)$$

where m is summed from 1 to M and j takes on any of the values from 1 to $\jmath$. We have these equations involving conditions on the coefficients u, insofar as they do not reduce merely to the constraint equations.

Let us look at these equations from the following point of view. Let us suppose that the u's are unknowns and that we have in (1-28) a number of non-homogeneous linear equations in these unknowns u, with coefficients which are functions of the q's and the p's. Let us look for a solution of these equations, which gives us the u's as functions of the q's and the p's, say

$$u_m = U_m(q, p). \qquad (1\text{-}29)$$

There must exist a solution of this type, because if there were none it would mean that the Lagrangian equations of motion are inconsistent, and we are excluding that case.

The solution is not unique. If we have one solution, we may add to it any solution $V_m(q, p)$ of the homogeneous equations associated with (1-28):

$$V_m[\phi_j, \phi_m] = 0, \qquad (1\text{-}30)$$

and that will give us another solution of the inhomogeneous equations (1-28). We want the most general solution of (1-28) and that means that we must consider *all* the independent solutions of (1-30), which we may denote by $V_{am}(q, p)$, $a = 1, \ldots, A$. The general solution of (1-28) is then

$$u_m = U_m + v_a V_{am}, \qquad (1\text{-}31)$$

in terms of coefficients v_a which can be arbitrary.

Let us substitute these expressions for u into the total Hamiltonian of the theory (1-22). That will give us the total Hamiltonian

$$H_T = H + U_m\phi_m + v_a V_{am}\phi_m. \qquad (1\text{-}32)$$

We can write this as

$$H_T = H' + v_a\phi_a, \qquad (1\text{-}33)$$

where $\qquad\qquad H' = H + U_m\phi_m \qquad (1\text{-}33)'$

and $\qquad\qquad\qquad \phi_a = V_{am}\phi_m. \qquad (1\text{-}34)$

In terms of this total Hamiltonian (1-33) we still have the equations of motion (1-21).

As a result of carrying out this analysis, we have satisfied all the consistency requirements of the theory and we still have arbitrary coefficients v. The number of the coefficients v will usually be less than the number of coefficients u. The u's are not arbitrary but have to satisfy consistency conditions, while the v's are arbitrary coefficients. We may take the v's to be arbitrary functions of the time and we have still satisfied all the requirements of our dynamical theory.

This provides a difference of the generalized Hamiltonian formalism from what one is familiar with in elementary dynamics. We have arbitrary functions of the time occurring in the general solution of the equations of motion with given initial conditions. These arbitrary functions of the time must mean that we are using a mathematical framework containing arbitrary features, for example, a coordinate system which we can choose in some arbitrary way, or the gauge in electrodynamics. As a result of this arbitrariness in the mathematical framework, the dynamical variables at future times are not completely determined by the initial dynamical variables, and this shows itself up through arbitrary functions appearing in the general solution.

We require some terminology which will enable one to appreciate the relationships between the quantities which occur in the formalism. I find the following terminology useful. I define any dynamical variable, R, a function of the q's and the p's, to be *first-class* if it has zero Poisson brackets with all the ϕ's:

$$[R, \phi_j] \approx 0, \quad j = 1, \ldots, J. \qquad (1\text{-}35)$$

It is sufficient if these conditions hold weakly. Otherwise R is *second-class*. If R is *first-class*, then $[R, \phi_j]$ has to be strongly equal to some linear function of the ϕ's, as anything that is weakly zero in the present theory is strongly equal to some linear function of the ϕ's. The ϕ's are, by definition, the only independent quantities which are weakly zero. So we have the strong equations

$$[R, \phi_j] = r_{jj'}\phi_{j'}. \qquad (1\text{-}36)$$

Before going on, I would like to prove a

Theorem: the Poisson bracket of two first-class quantities is also first-class. *Proof.* Let R, S be first-class: then in addition to (1-36), we have

$$[S, \phi_j] = s_{jj'}\phi_{j'}. \qquad (1\text{-}36)'$$

Let us form $[[R, S], \phi_j]$. We can work out this Poisson bracket using Jacobi's identity (1-13)

$$
\begin{aligned}
[[R, S], \phi_j] &= [[R, \phi_j], S] - [[S, \phi_j], R] \\
&= [r_{jj'}\phi_{j'}, S] - [s_{jj'}\phi_{j'}, R] \\
&= r_{jj'}[\phi_{j'}, S] + [r_{jj'}, S]\phi_{j'} - s_{jj'}[\phi_{j'}, R] \\
&\quad - [s_{jj'}, R]\phi_{j'} \\
&\approx 0
\end{aligned}
$$

by (1-36), (1-36)', the product law (1-12), and (1-20). The whole thing vanishes weakly. We have proved therefore that $[R, S]$ is first-class.

We have altogether four different kinds of constraints. We can divide constraints into first-class and second-class, which is quite independent of the division into primary and secondary.

I would like you to notice that H' given by (1-33)' and

the ϕ_a given by (1-34) are first-class. Forming the Poisson bracket of ϕ_a with ϕ_j we get, by (1-34), $V_{am}[\phi_m, \phi_j]$ plus terms that vanish weakly. Since the V_{am} are defined to satisfy (1-30), ϕ_a is first-class. Similarly (1-28) with U_m for u_m shows that H' is first-class. Thus (1-33) gives the total Hamiltonian in terms of a first-class Hamiltonian H' together with some first-class ϕ's.

Any linear combination of the ϕ's is of course another constraint, and if we take a linear combination of the primary constraints we get another primary constraint. So each ϕ_a is a primary constraint; and it is first-class. So the final situation is that we have the total Hamiltonian expressed as the sum of a first-class Hamiltonian plus a linear combination of the primary, first-class constraints.

The number of independent arbitrary functions of the time occurring in the general solution of the equations of motion is equal to the number of values which the suffix a takes on. That is equal to the number of independent primary first-class constraints, because all the independent primary first-class constraints are included in the sum (1-33).

That gives you then the general situation. We have deduced it by just starting from the Lagrangian equations of motion, passing to the Hamiltonian and working out consistency conditions.

From the practical point of view one can tell from the general transformation properties of the action integral what arbitrary functions of the time will occur in the general solution of the equations of motion. To each of these functions of the time there must correspond some primary first-class constraint. So we can tell which primary first-class constraints we are going to have without going through all the detailed calculation of working out Poisson brackets; in practical applications of this theory we can obviously save a lot of work by using that method.

I would like to go on a bit more and develop one further point of the theory. Let us try to get a physical understanding of the situation where we start with given initial variables and get a solution of the equations of motion containing arbitrary functions. The initial variables which we need are the q's and the p's. We don't need to be given initial values for the coefficients v. These initial conditions describe what physicists would call the *initial physical state* of the system. The physical state is determined only by the q's and the p's and not by the coefficients v.

Now the initial state must determine the state at later times. But the q's and the p's at later times are not uniquely determined by the initial state because we have the arbitrary functions v coming in. That means that the state does not uniquely determine a set of q's and p's, even though a set of q's and p's uniquely determines a state. There must be several choices of q's and p's which correspond to the same state. So we have the problem of looking for all the sets of q's and p's that correspond to one particular physical state.

All those values for the q's and p's at a certain time which can evolve from one initial state must correspond to the same physical state at that time. Let us take particular initial values for the q's and the p's at time $t = 0$, and consider what the q's and the p's are after a short time interval δt. For a general dynamical variable g, with initial value g_0, its value at time δt is

$$g(\delta t) = g_0 + \dot{g}\,\delta t$$
$$= g_0 + [g, H_T]\,\delta t$$
$$= g_0 + \delta t\{[g, H'] + v_a[g, \phi_a]\}. \quad (1\text{-}37)$$

The coefficients v are completely arbitrary and at our disposal. Suppose we take different values, v', for these coefficients. That would give a different $g(\delta t)$, the difference being

$$\Delta g(\delta t) = \delta t(v_a - v'_a)[g, \phi_a]. \quad (1\text{-}38)$$

We may write this as

$$\Delta g(\delta t) = \varepsilon_a[g, \phi_a], \quad (1\text{-}39)$$

where $\qquad \varepsilon_a = \delta t(v_a - v'_a) \qquad (1\text{-}40)$

is a small arbitrary number, small because of the coefficient δt and arbitrary because the v's and the v''s are arbitrary. We can change all our Hamiltonian variables in accordance with the rule (1-39) and the new Hamiltonian variables will describe the same state. This change in the Hamiltonian variables consists in applying an infinitesimal contact transformation with a generating function $\varepsilon_a\phi_a$. We come to the conclusion that the ϕ_a's, which appeared in the theory in the first place as the primary first-class constraints, have this meaning: *as generating functions of infinitesimal contact transformations, they lead to changes in the q's and the p's that do not affect the physical state.*

However, that is not the end of the story. We can go on further in the same direction. Suppose we apply two of these contact transformations in succession. Apply first a contact transformation with generating function

$\varepsilon_a \phi_a$ and then apply a second contact transformation with generating function $\gamma_{a'} \phi_{a'}$, where the gamma's are some new small coefficients. We get finally

$$g' = g_0 + \varepsilon_a[g, \phi_a] + \gamma_{a'}[g + \varepsilon_a[g, \phi_a], \phi_{a'}]. \quad (1\text{-}41)$$

(I retain the second order terms involving products $\varepsilon\gamma$, but I neglect the second order terms involving ε^2 or involving γ^2. This is legitimate and sufficient. I do that because I do not want to write down more than I really need for getting the desired result.) If we apply the two transformations in succession in the reverse order, we get finally

$$g'' = g_0 + \gamma_{a'}[g, \phi_{a'}] + \varepsilon_a[g + \gamma_{a'}[g, \phi_{a'}], \phi_a]. \quad (1\text{-}42)$$

Now let us subtract these two. The difference is

$$\Delta g = \varepsilon_a \gamma_{a'} \left\{ [[g, \phi_a], \phi_{a'}] - [[g, \phi_{a'}], \phi_a] \right\}. \quad (1\text{-}43)$$

By Jacobi's identity (1-13) this reduces to

$$\Delta g = \varepsilon_a \gamma_{a'} [g, [\phi_a, \phi_{a'}]]. \quad (1\text{-}44)$$

This Δg must also correspond to a change in the q's and the p's which does not involve any change in the physical state, because it is made up by processes which individually don't involve any change in the physical state. Thus we see that we can use

$$[\phi_a, \phi_{a'}] \quad (1\text{-}45)$$

as a generating function of an infinitesimal contact transformation and it will still cause no change in the physical state.

Now the ϕ_a are first-class: their Poisson brackets are weakly zero, and therefore strongly equal to some linear function of the ϕ's. This linear function of the ϕ's must be first-class because of the theorem I proved a little while back, that the Poisson bracket of two first-class quantities is first-class. So we see that the transformations which we get this way, corresponding to no change in the physical state, are transformations for which the generating function is a first-class constraint. The only way these transformations are more general than the ones we had before is that the generating functions which we had before are restricted to be first-class primary constraints. Those that we get now could be first-class secondary constraints. The result of this calculation is to show that we might have a first-class secondary constraint as a generating function of an infinitesimal contact transformation which leads to a change in the q's and the p's without changing the state.

For the sake of completeness, there is a little bit of further work one ought to do which shows that a Poisson bracket $[H', \phi_a]$ of the first-class Hamiltonian H' with a first-class ϕ is again a linear function of first-class constraints. This can also be shown to be a possible generator for infinitesimal contact transformations which do not change the state.

The final result is that those transformations of the dynamical variables which do not change physical states are infinitesimal contact transformations in which the generating function is a primary first-class constraint or possibly a secondary first-class constraint. A good many of the secondary first-class constraints do turn up by the process (1-45) or as $[H', \phi_a]$. I think it may be that all the first-class secondary constraints should be included

[23]

among the transformations which don't change the physical state, but I haven't been able to prove it. Also, I haven't found any example for which there exist first-class secondary constraints which do generate a change in the physical state.

Lecture No. 2

THE PROBLEM OF QUANTIZATION

We were led to the idea that there are certain changes in the p's and q's that do not correspond to a change of state, and which have as generators first-class secondary constraints. That suggests that one should generalize the equations of motion in order to allow as variation of a dynamical variable g with the time not only any variation given by (1-21), but also any variation which does not correspond to a change of state. So we should consider a more general equation of motion

$$\dot{g} = [g, H_E] \qquad (2\text{-}1)$$

with an extended Hamiltonian H_E, consisting of the previous Hamiltonian, H_T, plus all those generators which do not change the state, with arbitrary coefficients:

$$H_E = H_T + v'_{a'}\phi_{a'}. \qquad (2\text{-}2)$$

Those generators $\phi_{a'}$, which are not included already in H_T will be the first-class secondary constraints. The presence of these further terms in the Hamiltonian will give further changes in g, but these further changes in g do not correspond to any change of state and so they should certainly be included, even though we did not arrive at these further changes of *g* by direct work from the Lagrangian.

That, then, is the general Hamiltonian theory. The theory as I have developed it applies to a finite number of degrees of freedom but we can easily extend it to the case of an infinite number of degrees of freedom. Our suffix denoting the degree of freedom is $n = 1, \ldots, N$; we may easily make N infinite. We may further generalize it by allowing the number of degrees of freedom to be continuously infinite. That is to say, we may have as our q's and p's variables q_x, p_x where x is a suffix which can take on all values in a continuous range. If we work with this continuous x, then we have to change all our sums over n in the previous work into integrals. The previous work can all be taken over directly with this change.

There is just one equation which we will have to think of a bit differently, the equation which defines the momentum variables,

$$p_n = \frac{dL}{d\dot{q}_n}. \tag{1-3}$$

If n takes on a continuous range of values, we have to understand by this partial differentiation a process of partial functional differentiation that can be made precise in this way: We vary the velocities by $\delta\dot{q}_x$ in the Lagrangian and then put

$$\delta L = \int p_x \, \delta\dot{q}_x. \tag{2-3}$$

The coefficient of $\delta\dot{q}_x$ occurring in the integrand in δL is defined to be p_x.

After giving this general abstract theory, I think it would be a help if I gave a simple example as illustration.

I will take as an example just the electromagnetic field of Maxwell, which is defined in terms of potentials A_μ. The dynamical coordinates now consist of the potentials for all points of space at a certain time. That is to say, the dynamical coordinates consist of $A_{\mu x}$, where the suffix x stands for the three coordinates x^1, x^2, x^3 of a point in three-dimensional space at a certain time x^0 (not the four x's which one is used to in relativity). We shall have then as the dynamical velocities the time derivatives of the dynamical coordinates, and I shall denote these by a suffix 0 preceded by a comma.

Any suffix with a comma before it denotes differentiation according to the general scheme

$$\xi_{,\mu} = \frac{d\xi}{dx^\mu}. \tag{2-4}$$

We are dealing with special relativity so that we can raise and lower these suffixes according to the rules of special relativity: we have a change in sign if we raise or lower a suffix 1, 2, or 3 but no change of sign when we raise or lower the suffix 0.

We have as our Lagrangian for the Maxwell electrodynamics, if we work in Heaviside units,

$$L = -\frac{1}{4} \int F_{\mu\nu} F^{\mu\nu} \, d^3x. \tag{2-5}$$

Here d^3x means $dx^1 \, dx^2 \, dx^3$, the integration is over three-dimensional space, and $F_{\mu\nu}$ means the field quantities defined in terms of the potentials by

$$F_{\mu\nu} = A_{\nu,\mu} - A_{\mu,\nu}. \tag{2-6}$$

This L is the Lagrangian because its time integral is the action integral of the Maxwell field.

[27]

Let us now take this Lagrangian and apply the rules of our formalism for passing to the Hamiltonian. We first of all have to introduce the momenta. We do that by varying the velocities in the Lagrangian. If we vary the velocities, we have

$$\delta L = -\frac{1}{2} \int F^{\mu\nu} \, \delta F_{\mu\nu} \, d^3x$$

$$= \int F^{\mu 0} \, \delta A_{\mu,0} \, d^3x. \tag{2-7}$$

Now the momenta B^μ are defined by

$$\delta L = \int B^\mu \, \delta A_{\mu 0} \, d^3x \tag{2-8}$$

and these momenta will satisfy the basic Poisson bracket relations

$$[A_{\mu x}, B^\nu_{x'}] = g^\nu_\mu \, \delta^3(x - x'); \quad \mu, \nu = 0, 1, 2, 3. \tag{2-9}$$

In this formula A is taken at a point x in three-dimensional space and B is taken at a point x' in the three-dimensional space. g^ν_μ is just the Kronecker delta function. $\delta^3(x - x')$ is the three-dimensional delta function of $x - x'$.

We compare the two expressions (2-7) and (2-8) for δL and that gives us

$$B^\mu = F^{\mu 0}. \tag{2-10}$$

Now $F^{\mu\nu}$ is anti-symmetrical

$$F^{\mu\nu} = -F^{\nu\mu}. \tag{2-11}$$

So if we put $\mu = 0$, in (2-10) we get zero. Thus B^0_x is

[28]

equal to zero. This is a primary constraint. I write it as a weak equation:

$$B_x^0 \approx 0. \qquad (2\text{-}12)$$

The other three momenta $B^r (r = 1, 2, 3)$ are just equal to the components of the electric field.

I should remind you that (2-12) is not just one primary constraint: there is a whole threefold infinity of primary constraints because there is the suffix x which stands for some point in three-dimensional space; and each value for x will give us a different primary constraint.

Let us now introduce the Hamiltonian. We define that in the usual way by

$$
\begin{aligned}
H &= \int B^\mu A_{\mu,0} \, d^3x - L \\
&= \int (F^{r0} A_{r,0} + \tfrac{1}{4} F^{rs} F_{rs} + \tfrac{1}{2} F^{r0} F_{r0}) \, d^3x \\
&= \int (\tfrac{1}{4} F^{rs} F_{rs} - \tfrac{1}{2} F^{r0} F_{r0} + F^{r0} A_{0,r}) \, d^3x \\
&= \int (\tfrac{1}{4} F^{rs} F_{rs} + \tfrac{1}{2} B^r B^r - A_0 B^r_{,r}) \, d^3x. \qquad (2\text{-}13)
\end{aligned}
$$

I've done a partial integration of the last term in (2-13) to get it in this form. Now here we have an expression for the Hamiltonian which does not involve any velocities. It involves only dynamical coordinates and momenta. It is true that F_{rs} involves partial differentiations of the potentials, but it involves partial differentiations only with respect to x^1, x^2, x^3. That does not bring in any velocities. These partial derivatives are functions of the dynamical coordinates.

We can now work out the consistency conditions by

using the primary constraints (2-12). Since they have to remain satisfied at all times, $[B^0, H]$ has to be zero. This leads to the equation

$$B^r_{,r} \approx 0. \qquad (2\text{-}14)$$

This is again a constraint because there are no velocities occurring in it. This is a secondary constraint, which appears in the Maxwell theory in this way. If we proceed further to examine the consistency relations, we must work out

$$[B^r_{,r}, H] = 0. \qquad (2\text{-}15)$$

We find that this reduces to $0 = 0$. It does not give us anything new, but is automatically satisfied. We have therefore obtained all the constraints in our problem. (2-12) gives the primary constraints. (2-14) gives the secondary constraints.

We now have to look to see whether they are first-class or second-class, and we easily see that they are all first-class. The B_0 are momenta variables. They all have zero Poisson brackets with each other. $B^r_{,r}$ and B_0 also have zero Poisson brackets with each other. And $B^r_{,rx}$ and $B^r_{,rx'}$ also have zero Poisson brackets with each other. All these quantities are therefore first-class constraints. There are no second-class constraints occurring in the Maxwell electrodynamics.

The expression (2-13) for H is first-class, so this H can be taken as the H' of (1-33). Let us now see what the total Hamiltonian is:

$$H_T = \int (\tfrac{1}{4} F^{rs} F_{rs} + \tfrac{1}{2} B_r B_r) \, d^3x - \int A_0 B^r_{,r} \, d^3x$$
$$+ \int v_x B^0 \, d^3x. \qquad (2\text{-}16)$$

This v_x is an arbitrary coefficient for each point in three-dimensional space. We have just added on the primary first-class constraints with arbitrary coefficients, which is what we must do according to the rules to get the total Hamiltonian.

In terms of the total Hamiltonian we have the equation of motion in the standard form

$$\dot{g} \approx [g, H_T]. \qquad (1\text{-}21)$$

The g which we have here may be any field quantity at some point x in three-dimensional space, or may also be a function of field quantities at *different* points in three-dimensional space. It could, for example, be an integral over three-dimensional space. This g can be perfectly generally any function of the q's and the p's throughout three-dimensional space.

It is permissible to take $g = A_0$ and then we get

$$A_{0,0} = v, \qquad (2\text{-}17)$$

because A_0 has zero Poisson brackets with everything except the B_0 occurring in the last term of (2-16). This gives us a meaning for the arbitrary coefficient v_x occurring in the total Hamiltonian. It is the time derivative of A_0.

Now to get the most general motion which is physically permissible, we ought to pass over to the extended Hamiltonian. To do this we add on the first-class secondary constraints with arbitrary coefficients u_x. This gives the extended Hamiltonian:

$$H_E = H_T + \int u_x B^r_{,r} \, d^3x. \qquad (2\text{-}18)$$

Bringing in this extra term into the Hamiltonian allows

a more general motion. It gives more variation of the q's and the p's, of the nature of a gauge transformation. When this additional variation of the q's and the p's is brought in, it leads to a further set of q's and p's which must correspond to the same state.

That is the result of working out, according to our rules, the Hamiltonian form of the Maxwell theory. When we've got to this stage, we see that there is a certain simplification which is possible. This simplification comes about because the variables A_0, B_0 are not of any physical significance. Let us see what the equations of motion tell us about A_0 and B_0. $B_0 = 0$ all the time. That is not of interest. A_0 is something whose time derivative is quite arbitrary. That again is something which is not of interest. The variables A_0 and B_0 are therefore not of interest at all. We can drop them out from the theory and that will lead to a simplified Hamiltonian formalism where we have fewer degrees of freedom, but still retain all the degrees of freedom which are physically of interest.

In order to carry out this discard of the variables A_0 and B_0, we drop out the term $v_x B^0$ from the Hamiltonian. This term merely has the effect of allowing A_0 to vary arbitrarily. The term $-A_0 B^r_{,r}$ in H_T can be combined with the $u_x B^r_{,r}$ in the extended Hamiltonian. The coefficient u_x is an arbitrary coefficient in any case. When we combine these two terms, we just have this u_x replaced by $u'_x = u_x - A_0$ which is equally arbitrary. So that we get a new Hamiltonian

$$H = \int (\tfrac{1}{4}F^{rs}F_{rs} + \tfrac{1}{2}B_r B_r)\, d^3x + \int u'_x\, B^r_{,r}\, d^3x.$$

$$(2\text{-}19)$$

This Hamiltonian is sufficient to give the equations of motion for all the variables which are of physical interest. The variables A_0, B_0 no longer appear in it. This is the Hamiltonian for the Maxwell theory in its simplest form.

Now the usual Hamiltonian which people work with in quantum electrodynamics is not quite the same as that. The usual one is based on a theory which was originally set up by Fermi. Fermi's theory involves putting this restriction on the potentials:

$$A^\mu_{,\mu} = 0. \qquad (2\text{-}20)$$

It is quite permissible to bring in this restriction on the gauge. The Hamiltonian theory which I have given here does not involve this restriction, so that it allows a completely general gauge. It's thus a somewhat different formalism from the Fermi formalism. It's a formalism which displays the full transforming power of the Maxwell theory, which we get when we have completely general changes of gauge. This Maxwell theory gives us an illustration of the general ideas of primary and secondary constraints.

I would like now to go back to general theory and to consider the problem of quantizing the Hamiltonian theory. To discuss this question of quantization, let us first take the case when there are no second-class constraints, when all the constraints are first-class. We make our dynamical coordinates and momenta, the q's and p's, into operators satisfying commutation relations which correspond to the Poisson bracket relations of the classical theory. That is quite straightforward. Then we set up a Schrödinger equation

$$i\hbar \frac{d\psi}{dt} = H'\psi. \qquad (2\text{-}21)$$

ψ is the wave function on which the q's and the p's operate. H' is the first-class Hamiltonian of our theory.

We further impose certain supplementary conditions on the wave function, namely:

$$\phi_j \psi = 0. \tag{2-22}$$

Each of our constraints thus leads to a supplementary condition on the wave function. (The constraints, remember, are now all first-class.)

The first thing we have to do now is to see whether these equations for ψ are consistent with one another. Let us take two of the supplementary conditions and see whether they are consistent. Let us take (2-22) and

$$\phi_{j'} \psi = 0. \tag{2-22}'$$

If we multiply (2-22) by $\phi_{j'}$, we get

$$\phi_{j'} \phi_j \psi = 0. \tag{2-23}$$

If we multiply (2-22)' by ϕ_j, we get

$$\phi_j \phi_{j'} \psi = 0. \tag{2-23}'$$

If we now subtract these two equations, we get:

$$[\phi_j, \phi_{j'}]\psi = 0. \tag{2-24}$$

This further condition on ψ is necessary for consistency. Now we don't want to have any fresh conditions on ψ. We want all the conditions on ψ to be included among (2-22). That means to say, we want to have (2-24) a consequence of (2-22) which means we require

$$[\phi_j, \phi_{j'}] = c_{jj'j''}\phi_{j''}. \tag{2-25}$$

[34]

If (2-25) *does* hold, then (2-24) is a consequence of (2-22) and is not a new condition on the wave function.

Now we know that the ϕ's are all first-class in the classical theory, and that means that the Poisson bracket of any two of the ϕ's is a linear combination of the ϕ's in the classical theory. When we go over to the quantum theory, we must have a similar equation holding for the commutator, but it does not necessarily follow that the coefficients c are all on the left. We need to have these coefficients all on the left, because the c's will in general be functions of the coordinates and momenta and will not commute with the ϕ's in the quantum theory, and (2-24) will be a consequence of (2-22) only provided the c's are all on the left.

When we set up the quantities ϕ in the quantum theory, there may be some arbitrariness coming in. The corresponding classical expressions may involve quantities which don't commute in the quantum theory and then we have to decide on the order in which to put the factors in the quantum theory. We have to try to arrange the order of these factors so that we have (2-25) holding with all the coefficients on the left. If we can do that, then we have the supplementary conditions all consistent with each other. If we cannot do it, then we are out of luck and we cannot make an accurate quantum theory. In any case we have a first approximation to the quantum theory, because our equations would be all right if we look at them only to the order of accuracy of Planck's constant $\hbar$ and neglect quantities of order $\hbar^2$.

I have just discussed the requirements for the supplementary conditions to be consistent with one another. There is a similar discussion needed in order to check that the supplementary conditions shall be consistent

[35]

with the Schrödinger equation. If we start with a ψ satisfying the supplementary conditions (2-24) and let that ψ vary with the time in accordance with the Schrödinger equation, then after a lapse of a short interval of time will our ψ still satisfy the supplementary conditions? We can work out the requirement for that to be the case and we get

$$[\phi_j, H]\psi = 0, \qquad (2\text{-}26)$$

which means that $[\phi_j, H]$ must be some linear function of the ϕ's:

$$[\phi_j, H] = b_{jj'}\phi_{j'}, \qquad (2\text{-}27)$$

if we are not to get a new supplementary condition. Again we have an equation which we know is all right in the classical theory. ϕ_j and H are both first-class, so their Poisson bracket vanishes weakly. The Poisson bracket is thus strongly equal to some linear function of the ϕ's in the classical theory. Again we have to try to arrange things so that in the corresponding quantum equation we have all our coefficients on the left. That is necessary to get an accurate quantum theory, and we need a bit of luck, in general, in order to be able to bring it about.

Let us now consider how to quantize a Hamiltonian theory in which there are second-class constraints. Let us think of this question first in terms of a simple example. We might take as the simplest example of two second-class constraints

$$q_1 \approx 0 \quad \text{and} \quad p_1 \approx 0. \qquad (2\text{-}28)$$

If we have these two constraints appearing in the theory, then their Poisson bracket is not zero, so they are second-class. What can we do with them when we go over to the quantum theory? We cannot impose (2-28) as supplementary conditions on the wave function as we did with the first-class constraints. If we try to put $q_1\psi = 0$, $p_1\psi = 0$, then we should immediately get a contradiction because we should have $(q_1 p_1 - p_1 q_1)\psi = i\hbar\psi = 0$. So that won't do. We must adopt some different plan.

Now in this simple case it's pretty obvious what the plan must be. The variables q_1 and p_1 are not of interest if they are both restricted to be zero. So the degree of freedom 1 is not of any importance. We can just discard the degree of freedom 1 and work with the other degrees of freedom. That means a different definition for a Poisson bracket. We should have to work with a definition of a Poisson bracket in the classical theory

$$[\xi, \eta] = \frac{\partial \xi}{\partial q_n}\frac{\partial \eta}{\partial p_n} - \frac{\partial \xi}{\partial p_n}\frac{\partial \eta}{\partial q_n} \qquad \text{summed over } n = 2, \ldots N.$$

(2-29)

This would be sufficient because it would deal with all the variables which are of physical interest. Then we could just take q_1 and p_1 as identically zero. There's no contradiction involved there, and we can pass over to the quantum theory, setting it up in terms only of the degrees of freedom $n = 2, \ldots, N$.

In this simple case it is fairly obvious what we have to do to build up a quantum theory. Let us try now to generalize it. Suppose we have $p_1 \approx 0$, $q_1 \approx f(q_r, p_r)$, $r = 2, \ldots, N$, so f is any function of all the other q's and p's. We could drop out the number 1 degree of freedom if we substitute $f(q_r, p_r)$ for q_1 in the Hamiltonian and in all the other constraints. Again we can forget about the number 1 degree of freedom and simply work with the other degrees of freedom and pass over to a quantum theory in these other degrees of freedom. Again we should have to work with the (2-29) kind of Poisson bracket, referring only to the other degrees of freedom.

That is the idea which one uses for quantizing a theory which involves second-class constraints. The existence of second-class constraints means that there are some degrees of freedom which are not physically important. We have to pick out these degrees of freedom and set up new Poisson brackets referring only to the other degrees of freedom which *are* of physical importance. Then in terms of those new Poisson brackets we can pass over to the quantum theory. I would like to discuss a general procedure for carrying that out.

For the present, we are going back to the classical theory. We have a number of constraints $\phi_j \approx 0$, some of them first-class, some second-class. We can replace these constraints by independent linear combinations of them, which will do just as well as the original constraints. We try to arrange to take the linear combinations in such a way as to have as many constraints as possible brought into the first class. There may then be some left in the second class which we just cannot bring into the first class by taking linear combinations of them. Those which are left in the second class I will call χ_s, $s = 1, \ldots, S$. S is the number of second-class constraints which are such that no linear combination of them is first-class.

We take these surviving second-class constraints and we form all their Poisson brackets with each other and arrange these Poisson brackets as a determinant Δ:

$$\Delta = \begin{vmatrix} 0 & [\chi_1, \chi_2] & [\chi_1, \chi_3] & \cdots & [\chi_1, \chi_s] \\ [\chi_2, \chi_1] & 0 & [\chi_2, \chi_3] & \cdots & [\chi_2, \chi_s] \\ \vdots & \vdots & \vdots & & \vdots \\ [\chi_s, \chi_1] & [\chi_s, \chi_2] & [\chi_s, \chi_3] & \cdots & \vdots \end{vmatrix}$$

I would like now to prove a

Theorem: The determinant Δ does not vanish, not even weakly. *Proof*: Assume that the determinant *does* vanish. I'm going to show that we get a contradiction. If the determinant vanishes, then it is of some rank $T < S$. Now let us set up the determinant A:

$$A = \begin{vmatrix} \chi_1 & 0 & [\chi_1, \chi_2] & \cdots & [\chi_1, \chi_T] \\ \chi_2 & [\chi_2, \chi_1] & 0 & & [\chi_2, \chi_T] \\ \vdots & \vdots & \vdots & \vdots & \vdots \\ \chi_{T+1} & [\chi_{T+1}, \chi_1] & [\chi_{T+1}, \chi_2] & \cdots & [\chi_{T+1}, \chi_T] \end{vmatrix}$$

A has $T + 1$ rows and columns. $T + 1$ might equal S or might be less than S. If we expand A in terms of the elements of its first column, we will get each of these elements multiplied into one of the sub-determinants of Δ. Now I don't want all of these sub-determinants to vanish. It might so happen that they do all vanish. And in that case, I would choose the χ's which are referred to among the rows and columns of A in a different way. There must always be some way of choosing the χ's which occur in A so that the sub-determinants don't all vanish, because Δ is of rank T. So we choose the χ's in such a way that the coefficients of the elements in the first column are not all zero.

Now I will show that A has zero Poisson brackets with any of the ϕ's. If we form the Poisson bracket of ϕ with a determinant, we get the result by forming the Poisson bracket of ϕ with the first column of the determinant, adding on the result of forming the Poisson bracket of ϕ with the second column of the determinant, and so on. Thus

$$[\phi, A] = \begin{vmatrix} [\phi, \chi_1] & 0 & \cdots \\ [\phi, \chi_2] & [\chi_2, \chi_1] & \cdots \\ \vdots & \vdots & \\ [\phi, \chi_{T+1}] & [\chi_{T+1}, \chi_1] & \cdots \end{vmatrix}$$

$$+ \begin{vmatrix} \chi_1 & 0 & \cdots \\ \chi_2 & [\phi, [\chi_2, \chi_1]] & \cdots \\ \vdots & \vdots & \\ \chi_{T+1} & [\phi, [\chi_{T+1}, \chi_1]] & \cdots \end{vmatrix}$$

$$+ \begin{vmatrix} \chi_1 & 0 & [\phi, [\chi_1, \chi_2]] & \cdots \\ \chi_2 & [\chi_2, \chi_1] & 0 & \cdots \\ \vdots & \vdots & \vdots & \\ \chi_{T+1} & [\chi_{T+1}, \chi_1] & [\phi, [\chi_{T+1}, \chi_2]] & \cdots \end{vmatrix} + \cdots .$$

This looks rather complicated, but one can easily see that every one of these determinants vanishes. In the first place, the first determinant on the right vanishes: if ϕ is first class, then the first column vanishes; if ϕ is second class, then ϕ is one of the χ's and we have a determinant which is a part of the determinant Δ with $T + 1$ rows and columns. But Δ is assumed to be of rank T, so that any part of it with $T + 1$ rows and columns vanishes. Now, the second determinant on the right vanishes weakly because the first column vanishes weakly. Similarly all the other determinants vanish weakly. The result is that the whole right-hand side vanishes weakly.

[40]

Thus A is a quantity whose Poisson bracket with every one of the ϕ's vanishes weakly.

Also, we can expand the determinant A in terms of the elements of the first column, and get A as a linear combination of the χ's. So we have the result that a certain linear combination of the χ's has zero Poisson brackets with all the ϕ's. That means that this linear combination of the χ's is first class. That contradicts our assumption that we have put as many χ's as possible into the first class. That proves the theorem.

Incidentally, we see that the number of surviving χ's, which cannot be brought into the first class, must be even, because the determinant $\varDelta$ is antisymmetrical. Any antisymmetrical determinant with an odd number of rows and columns vanishes. This one doesn't vanish and therefore must have an even number of rows and columns.

Because this determinant, $\varDelta$, doesn't vanish, we can bring in the reciprocal $c_{ss'}$ of the matrix whose determinant is $\varDelta$. We define the matrix $c_{ss'}$ by

$$c_{ss'}[\chi_{s'}, \chi_{s''}] = \delta_{ss''}. \qquad (2\text{-}30)$$

We now define new Poisson brackets in accordance with this formalism: any two quantities ξ, η have a new Poisson bracket defined by

$$[\xi, \eta]^* = [\xi, \eta] - [\xi, \chi_s]c_{ss'}[\chi_{s'}, \eta]. \qquad (2\text{-}31)$$

It is easy to check that new Poisson brackets defined in this way satisfy the laws which Poisson brackets usually satisfy: $[\xi, \eta]^*$ is antisymmetrical between ξ and η, is linear in ξ, is linear in η, satisfies the product law $[\xi_1\xi_2, \eta]^* = \xi_1[\xi_2\eta]^* + [\xi_1, \eta]^*\xi_2$, and obeys the Jacobi identity $[[\xi, \eta]^*, \zeta]^* + [[\eta, \zeta]^*\xi]^* + [[\zeta, \xi]^*, \eta]^* = 0$.

I don't know of any neat way of proving the Jacobi identity for the new Poisson brackets. If one just substitutes according to the definition and works it out in a complicated way, one does find that all the terms cancel out and that the left-hand side equals zero. I think there ought to be some neat way of proving it, but I haven't been able to find it. The straightforward method I have given in the *Canadian Journal of Mathematics*, **2**, 147 (1950). The problem has been dealt with by Bergmann, *Physical Review*, **98**, 531 (1955).

Now let us see what we can do with these new Poisson brackets. First of all, I would like you to notice that the equations of motion are as valid for the new Poisson brackets as for the original ones.

$$[g, H_T]^* = [g, H_T] - [g, \chi_s]c_{ss'}[\chi_{s'}, H_T]$$
$$\approx [g, H_T]$$

because the terms $[\chi_{s'}, H_T]$ all vanish weakly on account of H_T being first-class. Thus we can write

$$\dot{g} \approx [g, H_T]^*.$$

Now if we take any function ξ whatever of the q's and p's, and form its new Poisson bracket with one of the χ's, say $\chi_{s''}$, we have

$$[\xi, \chi_{s''}]^* = [\xi, \chi_{s''}] - [\xi, \chi_s]c_{ss'}[\chi_{s'}, \chi_{s''}]$$
$$= [\xi, \chi_{s''}] - [\xi, \chi_s]\delta_{ss''} \qquad \text{by (2-30)}$$
$$= 0.$$

Thus we can put the χ's equal to 0 before working out new Poisson brackets. This means that the equation

$$\chi_s = 0 \qquad (2\text{-}32)$$

may be considered as a strong equation.

We modify our classical theory in this way, bringing in these new Poisson brackets, and this prepares the ground for passing to the quantum theory. We pass over to the quantum theory by taking the commutation relations to correspond to the new Poisson bracket relations and taking the strong equations (2-32) to be equations between operators in the quantum theory. The remaining weak equations, which are all first class, become again supplementary conditions on the wave functions. The situation is then reduced to the previous case where there were only first-class ϕ's. We have again, therefore, a method of quantizing our general classical Hamiltonian theory. Of course, we again need a bit of luck in order to arrange that the coefficients are all on the left in the consistency conditions.

That gives the general method of quantization. You notice that when we have passed over to the quantum theory, the distinction between primary constraints and secondary constraints ceases to be of any importance. The distinction between primary and secondary constraints is not a very fundamental one. It depends very much on the original Lagrangian which we start off with. Once we have gone over to the Hamiltonian formalism, we can really forget about the distinction between primary and secondary constraints. The distinction between first-class and second-class constraints is very important. We must put as many as possible into the first class and bring in new Poisson brackets which enable us to treat the surviving second-class constraints as strong.

Lecture No. 3

QUANTIZATION ON CURVED SURFACES

We started off with a classical action principle. We took our action integral to be Lorentz-invariant. This action gives us a Lagrangian. We then passed from the Lagrangian to the Hamiltonian, and then to the quantum theory by following through certain rules. The result is that, starting with a classical field theory, described by an action principle, we end up with a quantum field theory. Now you might think that that finishes our work, but there is one important problem still to be considered: whether our quantum field theory obtained in this way is a relativistic theory. For the purposes of discussion, we may confine ourselves to special relativity. We have then to consider whether our quantum theory is in agreement with special relativity.

We started from an action principle and we required that our action should be Lorentz-invariant. That is sufficient to ensure that our classical theory shall be relativistic. The equations of motion that follow from a Lorentz invariant action principle must be relativistic equations. It is true that when we put these equations of motion into the Hamiltonian form, we are disturbing the four-dimensional symmetry. We are expressing our equations in the form

$$\dot{g} \approx [g, H_T]. \qquad (1\text{-}21)$$

The dot here means dg/dt and refers to one absolute time, so that the classical equations of motion in the Hamiltonian form are not manifestly relativistic, but we know that they must be relativistic in content because they follow from relativistic assumptions.

However, when we pass over to the quantum theory we are making new assumptions. The expression for H_T which we have in the classical theory does not uniquely determine the quantum Hamiltonian. We have to decide questions about the order in which to put non-commuting factors in the quantum theory. We have something at our disposal in choosing this order, and so we are making new assumptions. These new assumptions may disturb the relativistic invariance of the theory, so that the quantum field theory obtained by this method is not necessarily in agreement with relativity. We now have to face the problem of seeing how we can ensure that our quantum theory shall be a relativistic theory.

For that purpose we have to go back to first principles. It is no longer sufficient to consider just one time variable referring to one particular observer; we have to consider different observers moving relatively to one another. We must set up a quantum theory which applies equally to any of these observers, that is, to any time axis. To get a theory involving all the different time axes, we should first get the corresponding *classical* theory and then pass from this classical theory to the quantum theory by the standard rules.

I would like to go back to the beginning of our

Hamiltonian development and consider a special case. We started our development by taking a Lagrangian L, which is a function of dynamical coordinates and velocities q, $\dot{q}$, introducing the momenta, then introducing the Hamiltonian. Let us take the special case when L is homogeneous of the first degree in the $\dot{q}$'s. Then Euler's Theorem tells us that

$$\dot{q}_n \frac{\partial L}{\partial \dot{q}_n} = L. \tag{3-1}$$

That just tells us that $p_n \dot{q}_n - L = 0$. Thus we get in this special case a Hamiltonian that is zero.

We necessarily get primary constraints in this case. There must certainly be one primary constraint, because the p's are homogeneous functions of degree zero in the velocities. The p's are thus functions only of the ratios of the velocities. The number of p's is equal to N, the number of degrees of freedom, and the number of ratios of the velocities is $N - 1$. N functions of $N - 1$ ratios of the velocities cannot be independent. There must be at least one function of the p's and q's which is equal to zero; there must be at least one primary constraint. There may very well be more than one. One can also see that, if we are to have any motion at all with a zero Hamiltonian, we must have at least one primary first-class constraint.

We have the expression (1-33) for the total Hamiltonian

$$H_T = H' + v_a \phi_a.$$

H' must be a first-class Hamiltonian, and as 0 is certainly a first-class quantity we may take $H' = 0$. Our total

Hamiltonian is now built up entirely from the primary first-class constraints with arbitrary coefficients:

$$H_T = v_a \phi_a, \qquad (3\text{-}2)$$

showing that there must be at least one primary first-class constraint if we are to have any motion at all.

Our equations of motion now read like this:

$$\dot{g} \approx v_a[g, \phi_a].$$

We can see that the $\dot{g}$'s may all be multiplied by a factor because, since the coefficients v are arbitrary, we may multiply them all by a factor. If we multiply all the dg/dt's by a factor, it means that we have a different time scale. So we have now Hamiltonian equations of motion in which the time scale is arbitrary. We could introduce another time variable τ instead of t and use τ to give us equations of motion

$$\frac{dg}{d\tau} \approx v_a'[g, \phi_a]. \qquad (3\text{-}3)$$

So we have now a Hamiltonian scheme of equations of motion in which there is no absolute time variable. Any variable increasing monotonically with t could be used as time and the equations of motion would be of the same form. Thus the characteristic of a Hamiltonian theory where the Hamiltonian H' is zero and where every Hamiltonian is weakly equal to zero, is that there is no absolute time.

We may look at the question also from the point of view of the action principle. If I is the action integral, then

$$I = \int L(q, \dot{q}) \, dt = \int L\left(q, \frac{dq}{d\tau}\right) d\tau, \qquad (3\text{-}4)$$

because L is homogeneous of the first degree in the dq/dt. So we can express the action integral with respect to τ in the same form as with respect to t. That shows that the equations of motion which follow from the action principle must be invariant under the passage from t to τ. The equations of motion do not refer to any absolute time.

We have thus a special form of Hamiltonian theory, but in fact this form is not really so special because, starting with any Hamiltonian, it is always permissible to take the time variable as an extra coordinate and bring the theory into a form in which the Hamiltonian is weakly equal to zero. The general rule for doing this is the following: we take t and put it equal to another dynamical coordinate q_0. We set up a new Lagrangian

$$L^* = \frac{dq_0}{d\tau} L\left(q, \frac{dq/d\tau}{dq_0/d\tau}\right)$$
$$= L^*\left(q_k, \frac{dq_k}{d\tau}\right), \quad k = 0, 1, 2, \ldots, N \quad (3\text{-}5)$$

L^* involves one more degree of freedom than the original L. L^* is *not* equal to L but

$$\int L^* \, d\tau = \int L \, dt.$$

Thus the action is the same whether it refers to L^* and τ or to L and t. So for any dynamical system we can treat the time as an extra coordinate q_0 and then pass to a new Lagrangian L^*, involving one extra degree of freedom and homogeneous of the first degree in the velocities. L^* gives us a Hamiltonian which is weakly equal to zero.

This special case of the Hamiltonian formalism where

the Hamiltonian is weakly equal to zero is what we need for a relativistic theory, because in a relativistic theory we don't want to have one particular time playing a special role; we want to have the possibility of various times τ which are all on the same footing. Let us see in detail how we can apply this idea.

We want to consider states at specified times with respect to different observers. Now if we set up a space-time picture as in Fig. 1, the state at a certain time refers to the physical conditions on a three-dimensional flat space-like surface S_1 which is orthogonal to the time axis. The state at different times will refer to physical conditions on different surfaces S_2, S_3, ... Now we want to bring in other time axes referring to different observers and the state, with respect to the other time axes, will involve physical conditions on other flat space-like surfaces like S_1'. We want to have a Hamiltonian theory which will enable us to pass from the state, S_1 say, to the state S_1'. Starting off with given initial conditions on the surface S_1 and applying the equations of motion, we must be able to pass over to the physical conditions on the surface S_1'. There must thus be four freedoms in the motion of a state, one freedom corresponding to the movement of the surface parallel to itself, then three more freedoms corresponding to a general change of direction of this flat surface. That means that there will be four arbitrary functions occurring in the solution of the equations of motion which we are trying to get. So we need a Hamiltonian theory with (at least) four primary first-class constraints.

There may be other primary first-class constraints if there are other kinds of freedom in the motion, for example, if we have the possibility of the gauge

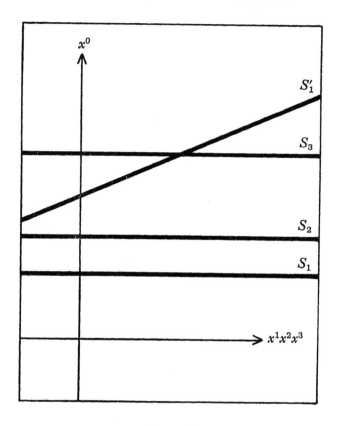

FIGURE 1

transformations of electrodynamics. To simplify the discussion, I will ignore this possibility of other first-class primary constraints, and consider only the ones which arise from the requirements of relativity.

We could proceed to set up our theory referring to these flat space-like surfaces which can move with the four freedoms, but I would like first to consider a more

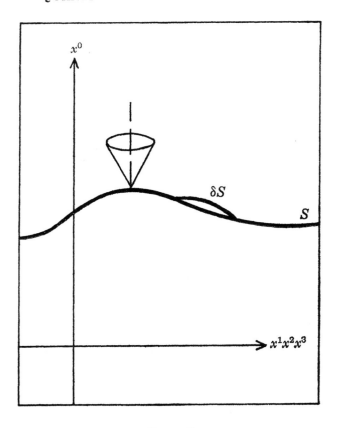

FIGURE 2

general theory in which we consider a state to be defined on an arbitrary curved space-like surface, such as S of Fig. 2. This represents a three-dimensional surface in space-time which has the property of being everywhere space-like, that is to say, the normal to the surface must lie within the light-cone. We may set up a Hamiltonian theory which tells us how the physical conditions vary when we go from one of the curved space-like surfaces to a neighboring one.

Now, bringing in the curved surfaces means bringing in something which is not necessary from the point of view of special relativity. If we wanted to bring in general relativity and gravitational fields, then it would be essential to work with these curved surfaces, but for special relativity, the curved surfaces are not essential. However, I like to bring them in at this stage, even for the discussion of a theory in special relativity, because I find it easier to explain the basic ideas with reference to these curved surfaces than with reference to the flat surfaces. The reason is that with these curved surfaces we can make local deformations of the surface like δS in Fig. 2, and discuss the equations of motion with respect to these local deformations of the surface.

One way of proceeding now would be to refer our action integral to a set of curved surfaces, like S, take the amount of action between two neighboring curved surfaces, divide it by some parameter $\delta \tau$ expressing the distance between these two surfaces, take this amount of action as our Lagrangian, then apply our standard method of passing from the Lagrangian to the Hamiltonian. Our Lagrangian would necessarily be homogeneous of the first degree in the velocities with respect to the time parameter τ which specifies the passage from one of these space-like surfaces to a neighboring one, and it would lead to a Hamiltonian theory for which the Hamiltonian is weakly equal to zero.

However, I don't want to go through all the work of following through in detail what we get from an action principle. I want to short-circuit that work and discuss the form of the final Hamiltonian theory which results.

[52]

We can get quite a lot of information about the form of this Hamiltonian theory just from our knowledge that there must be freedom for the space-like surface to move arbitrarily so long as it remains always space-like. This freedom of motion of the space-like surface must correspond to first-class primary constraints in the Hamiltonian, there being one primary first-class constraint for each type of elementary motion of the surface which can be set up. I shall develop the theory from that point of view.

First of all we have to introduce suitable dynamical variables. Let us describe a point on the space-like surface S by three curvilinear coordinates (x^1, x^2, x^3) $= (x^r)$. In order to fix the position of this space-like surface in space-time, we introduce another set of coordinates $y_A(A = 0, 1, 2, 3)$, which we may take to be rectilinear, orthogonal coordinates in special relativity. (I use a capital suffix for referring to the y coordinate system and a small suffix such as r for referring to the x coordinate system.) The four functions y_A, of x^r, will specify the surface S in space time and will also specify its parameterization, i.e. the system of coordinates x^1, x^2, x^3.

We can use these y_A as dynamical coordinates, q's. If we form

$$y_{A,r} = \frac{\partial y_A}{\partial x^r}, \quad (r = 1, 2, 3) \tag{3-6}$$

this is a function of the q's, the dynamical coordinates.

$$\dot{y}_A = \frac{\partial y_A}{\partial \tau}, \tag{3-7}$$

τ being the parameter changing from one surface to the

[53]

neighboring surface, will be a velocity, a $\dot{q}$. Thus y_A are the dynamical coordinates needed for describing the surface and $\dot{y}_A$ are the velocities.

We shall need to introduce momentum variables w_A conjugate to these dynamical coordinates. The momentum variables will be connected with the coordinates by the Poisson bracket relations

$$[y_{Ax}, w_{\Gamma x'}] = g_{A\Gamma}\delta^3(x - x'). \tag{3-8}$$

We shall need other variables for describing any physical fields which occur in the problem. If we are dealing with a scalar field V, then $V(x)$ for all values of x^1, x^2, x^3 will provide us with further dynamical coordinates, q's. V_r will be functions of the q's. $\partial V/\partial \tau$ will be a velocity. The derivative of V in any direction is expressible of terms of $\partial V/\partial \tau$ and V_r and so is expressible in terms of the dynamical coordinates and velocities. The Lagrangian will involve these V's differentiated in general directions and is thus a function of the dynamical coordinates and velocities. For each V, we shall need a conjugate momentum U, satisfying the Poisson bracket conditions

$$[V(x), U(x')] = \delta^3(x - x'). \tag{3-9}$$

That is how one would treat a scalar field. There is a similar method for vector, tensor, or spinor fields, just bringing in the necessary additional suffixes. I need not go into that.

Now let us see what the Hamiltonian will be like. The Hamiltonian has to be a linear function of primary first-class constraints of the type (3-2). First of all I shall put down what the primary first-class constraints are like. There must be primary first-class constraints which allow for arbitrary deformations of the surface. They must involve the variables w which are conjugate to the y's, in order to make the y's vary, and they will involve other field quantities. We can express them in the form

$$w_\Lambda + K_\Lambda \approx 0, \qquad (3\text{-}10)$$

where K_Λ is some function of the Hamiltonian variables, the q's and p's, not involving the w's.

Now we can assert that the Hamiltonian is just an arbitrary linear function of all the quantities (3-10):

$$H_T = \int c^\Lambda(w_\Lambda + K_\Lambda)\,d^3x. \qquad (3\text{-}11)$$

This is integrated over the three x's which specify a point on the surface. The c's are arbitrary functions of the three x's and the time.

The general equation of motion is of course $\dot{g} \approx [g, H_T]$. We can get a meaning for the coefficient c^Λ by taking this equation of motion and applying it for g equal to one of the y variables. For $g = y_\Lambda$ at some particular point x^1, x^2, x^3 we get

$$\dot{y}_\Lambda = \left[y_\Lambda, \int c'^\Gamma(w'_\Gamma + K'_\Gamma)\,d^3x' \right]$$

$$= \int c'^\Gamma[y_\Lambda, w'_\Gamma + K'_\Gamma]\,d^3x. \qquad (3\text{-}12)$$

Here the $'$ attached to a field quantity c^Γ, w_Γ or K_Γ denotes the value of that quantity at the point $X^{1'}$, $X^{2'}$, $X^{3'}$. y_Λ has zero Poisson brackets with K'_Γ because K'_Γ is independent of the w's, so we just have to take into account the Poisson bracket of y_A with $w'_\Gamma = w_\Gamma(x')$. This gives us the delta function and so

$$\dot{y}_A = c_A. \tag{3-13}$$

Thus the coefficients c_A turn out to be the velocity variables which tell us how our surface varies with the parameter τ. We can get an arbitrary variation of the surface with τ by choosing these c_A in an arbitrary way.

This tells us what the Hamiltonian is like for a field theory expressed with respect to states on curvilinear surfaces.

We can make a deeper analysis of this Hamiltonian by resolving the vectors which occur in it into components which are normal and tangential to the surface. If we have any vector whatever, ξ_A, we can obtain from ξ_A a *normal component*

$$\xi_\perp = \xi_A l^A,$$

where l^A is the unit normal vector, and *tangential components* (referred to the x coordinate system)

$$\xi_r = \xi_A y^A_{,r}.$$

The l are determined by the $y^A_{,r}$ and are thus functions of the dynamical coordinates. Any vector can be resolved in this way into a part normal to the surface and a part tangential to the surface. We have the scalar product law

$$\xi_A \eta^A = \xi_\perp \eta_\perp + \gamma^{rs} \xi_r \eta_s, \tag{3-14}$$

where $\gamma_{rs} \, dx^r \, dx^s$ is the metric in the surface referred to the x-coordinate system. γ^{rs} is the reciprocal matrix of the γ_{rs}. $(r, s = 1, 2, 3)$.

We can use this scalar product law (3-14) to express

[56]

our total Hamiltonian in terms of the tangential and normal components of w and K:

$$H_T = \int \dot{y}^A(w_A + K_A)\, d^3x$$

$$= \int \left(\dot{y}_\perp(w_\perp + K_\perp) + \gamma^{rs}\dot{y}_r(w_s + K_s) \right) d^3x.$$

(3-15)

Here $\dot{y} = \dot{y}^A l_A$ and $\dot{y}_r = \dot{y}^A y_{A,r}$.

We shall need the Poisson bracket relationships between the normal and tangential terms in (3-15). I will first write down the Poisson bracket relations for the different components of w. We have of course

$$[w_A, w'_\Gamma] = 0,$$

(3-16)

referred to the external coordinates y; but when we resolve our w's into normal and tangential components, they will no longer have zero Poisson brackets with each other. The Poisson brackets can easily be worked out by straightforward arguments. I don't want to go through the details of that work. I will just mention that the details can be found in a paper of mine (*Canadian Journal of Mathematics*, **3**, 1 (1951)). The results are

$$[w_r, w'_s] = w_s \delta_{,r}(x - x') + w'_r \delta_{,s}(x - x'), \qquad (3\text{-}17)$$

$$[w_\perp, w'_r] = w'_\perp \delta_{,r}(x - x'), \qquad (3\text{-}18)$$

$$[w_\perp, w'_\perp] = -2w^r \delta_{,r}(x - x') - w^r_{,r}\delta(x - x'), \quad (3\text{-}19)$$

Now we know that

$$[w_\mu + K_\mu, w'_\nu + K'_\nu] \approx 0 \quad \text{for } \mu, \nu = r, s \text{ or } \perp. \ (3\text{-}20)$$

We can infer that

$$[w_r + K_r, w'_s + K'_s]$$
$$= (w_s + K_s)\delta_{,r}(x - x') + (w'_r + K'_r)\delta_{,s}(x - x'),$$

$$(3\text{-}21)$$

$$[w_\perp + K_\perp, w'_r + K'_r]$$
$$= (w'_\perp + K'_\perp)\delta_{,r}(x - x'),$$

$$(3\text{-}22)$$

$$[w_\perp + K_\perp, w'_\perp + K'_\perp]$$
$$= -2(w^r + K^r)\delta_{,r}(x - x') - (w^r + K^r)_{,r}\delta(x - x').$$

$$(3\text{-}23)$$

These results could be worked out directly from the definitions of the normal and tangential components of the w's, but they can be inferred more simply by the following argument. Since $w_r + K_r$, $w_\perp + K_\perp$ are all first class, their Poisson brackets are zero weakly. Thus $[w_r + K_r, w'_s + K'_s]$, $[W_\perp + K_\perp, w'_r + K'_r]$ and $[w_\perp + K_\perp, w'_\perp + K'_\perp]$ must all be weakly equal to zero. We can now infer what they are equal to strongly. We have to put on the right-hand side in each of (3-21,) (3-22,) and (3-23) a quantity which is weakly equal to zero and which is therefore built up from $w_r + K_r$ and $w_\perp + K_\perp$ with certain coefficients. Further, we can see what these coefficients are by working out what terms containing w there are on the right-hand sides. Terms containing w's can arise only from taking the Poisson bracket of a w with a w, according to (3-17), (3-18), and (3-19). Taking a Poisson bracket $[w, K']$ will not lead to anything involving w, because it means taking the Poisson bracket of a w momentum with some functions of dynamical coordinates and momenta other than w's, and that won't involve the w momentum variables. Similarly the Poisson bracket of a K with a K won't involve any w variables. Thus the only w variables which occur on the right side of

(3-21) will be the ones which occur on the right side of (3-17). We have to put certain further terms in the right side of (3-21) in order that the total expression shall be weakly equal to zero. It is then quite clear what we should put here, namely, $(w_s + K_s)\delta_{,r}(x - x') + (w'_r + K'_r)\delta_{,s}(x - x')$. We do the same with the right sides of (3-22), (3-23).

The next thing to notice is that the terms $w_s + K_s$ in the Hamiltonian (3-15) correspond to a motion in which we change the system of coordinates in the curved surface but do not have the surface moving. It corresponds to each point in the surface moving tangentially to the surface.

Let us put $\dot{y}_\perp = 0$, which means that we are taking no motion of the surface perpendicular to itself but are merely making a change of the coordinates of the surface, and then we have equations of motion of the type

$$\dot{g} = \int \gamma^{rs} \dot{y}_r [g, w_s + K_s] \, d^3x. \qquad (3\text{-}24)$$

This must be the equation of motion which tells us how g varies when we change the system of coordinates in the surface without moving the surface itself. Now this change in g must be a trivial one, which can be inferred merely from the geometrical nature of the dynamical variable g. If g is a scalar, then we know how that changes when we change the system of coordinates x^1, x^2, x^3. If it is a component of a vector or a tensor there will be a rather more complicated change for g, but still we can work it out; similarly if g is a spinor. In every case, this change of g is a trivial one. That means that K_s can be determined from geometrical arguments only.

[59]

I will give one or two examples of that. For a scalar field V with a conjugate momentum U, there is a term

$$V_{,r}U \qquad (3\text{-}25)$$

in K_r. For a vector field, say a three-vector A_s, with conjugate B^s, there is a term

$$A_{s,r}B^s - (A_r B^s)_{,s} \qquad (3\text{-}26)$$

in K_r; and so on for tensors, with something rather more complicated for spinors. The first term in (3-26) is the change in A_s coming from the translation associated with the change in the system of coordinates, and the second is the change in the A_s arising from the rotation associated with the change in the system of coordinates. There is no such rotation term coming in in the case (3-25) of the scalar.

We can obtain the total K_r by adding the contribution needed for all the different kinds of fields which are present in the problem. The result is that we can work out this tangential component of K just from geometrical arguments. One can see in this way that the tangential component of K is something which is not of real physical importance, it is just concerned with the mathematical technique. The quantity which is of real physical importance is the normal component of K in (3-15). This normal component of K added on to the normal component of w gives us the first-class constraint which is associated with a motion of the surface normal to itself. That is something which is of dynamical importance.

The problem of getting a Hamiltonian field theory on these curved surfaces involves finding the expressions K to satisfy the required Poisson bracket relations

(3-21), (3-22), and (3-23). The tangential part of K can be worked out from geometrical arguments as I discussed, and when we have worked it out we should find of course that it satisfies the first Poisson bracket relation (3-21). The second Poisson bracket relation (3-22) involves $K_\perp$ linearly and this Poisson bracket relation would be satisfied by any quantity $K_\perp$ which satisfies the condition of being a scalar density. This Poisson bracket relation really tells us that if the quantity $K_\perp$ varies suitably under a change of coordinate system X^1, X^2, X^3, this Poisson bracket relation will be fulfilled. The difficult relation to fulfill is the third one, which is quadratic in $K_\perp$. So the problem of setting up a Hamiltonian field theory on curved space-like surfaces is reduced to the problem of finding a normal component of K which is a scalar density and which satisfies the Poisson bracket relationship (3-23).

One way of finding such a normal component of K is to work from a Lorentzinvariant action principle. We might obtain all the components of K by working from the action principle. If we did that, the tangential part of K which we get would not necessarily be the same as that built up from terms like (3-25) and (3-26), because it might differ by a contact transformation. But one could eliminate such a contact transformation by rewriting the action principle, adding to it a perfect differential term. This doesn't affect the equations of motion. By such a change of the action principle, one can arrange that the tangential part of K given by the action principle agrees precisely with the value which is obtained by the simple application of geometrical arguments. We are then able to find the normal component of K by working with our general method of passing from the action principle to the Hamiltonian. If the action principle is relativistic, then the normal component of K obtained in this way would have to satisfy the condition (3-23).

We can now discuss the passage to the quantum theory. Quantization involves making the quantities w and the variables which enter in K into operators. We have to be careful now how we define the tangential and the normal components of w, and I choose this way to define them:

$$w_r = y_{A,r} w^A, \qquad (3\text{-}27)$$

putting the momentum variable w on the right. (In the quantum theory, you see, the result is different, depending on whether we put the w on the right or the left.) Similarly,

$$w_\perp = l_A w^A. \qquad (3\text{-}28)$$

Then these quantities are well defined.

Now in the quantum theory we have the weak equations $w_r + K_r \approx 0$ and $w_\perp + K_\perp \approx 0$, which provide us with supplementary conditions on the wave function:

$$(w_r + K_r)\psi = 0, \qquad (3\text{-}29)$$
$$(w_\perp + K_\perp)\psi = 0, \qquad (3\text{-}30)$$

corresponding to (2-22). We require that these supplementary conditions be consistent. According to (2-25), we must arrange that in the commutation relations (3-21), (3-22), and (3-23) the coefficients on the right-hand sides stand before (on the left of) the constraints.

In the case of (3-21), the tangential components, the conditions fit if we choose the order of the factors in K_r so that the momentum variables are always on the right. We have now in (3-21) a number of quantities, linear in the momentum variables with the momentum variables on the right, and the commutator of any two such quantities will again be linear in the momentum variables with the momentum variables on the right. Thus we shall always have the momentum variables on the right and we shall always have our factors occurring in the order in which we want them to.

Now we have the problem of bringing in $K_\perp$, which cannot be disposed of so simply. $K_\perp$ will usually involve the product of non-commuting factors and we have to arrange the order of those factors so that (3-22) and (3-23) shall be satisfied with the coefficients occurring on the left in every term on the right-hand side. The equation (3-22) is again a fairly simple one to dispose of. If we simply take $K_\perp$ to be a scalar density, that is all that is needed, because we have $w_\perp + K_\perp$ occurring on the right-hand side without any coefficients which don't commute with it; the only coefficient is the delta function, which is a number.

But the relationship (3-23) is the troublesome one. For the purposes of the quantum theory, I ought to write out the right-hand side here rather more explicitly:

$$[w_\perp + K_\perp, w'_\perp + K'_\perp] = -2\gamma^{rs}(w_s + K_s)\delta_{,r}(x - x')$$
$$-\left(\gamma^{rs}(w_s + K_s)\right)_{,r}\delta(x - x'). \quad (3\text{-}31)$$

I've written this out with the coefficients γ^{rs} occurring on the left, and that is how we need to have these coefficients in the quantum theory.

The problem of setting up a quantum field theory on general curved surfaces involves finding $K_\perp$ so that this Poisson bracket relationship (3-31) holds with the coefficients γ^{rs} occurring on the left. If we do satisfy (3-31), then the supplementary conditions (3-30) are consistent with each other, and we already have (3-29) consistent with each other and (3-30) consistent with (3-29).

There we have formulated the conditions for our quantum theory to be relativistic. We need a bit of luck to be able to satisfy the conditions. We cannot always satisfy them. There is one general rule which is of importance, which tells us that when we've got a $K_\perp$ satisfying these conditions and certain other conditions, we can easily construct other $K_\perp$'s to satisfy the conditions. Let us suppose that we have a solution in which $K_\perp$ involves only undifferentiated momentum variables together with dynamical coordinates which may be differentiated. There are a number of simple fields for which $K_\perp$ does satisfy the Poisson bracket relations (3-22) and (3-23) and does have this simple character. Then we may add to $K_\perp$ any function of the undifferentiated q's. That is to say, we take a new $K_\perp$,

$$K_\perp^* = K_\perp + \phi(q).$$

Then we see that adding on this ϕ to $K_\perp$ can affect the right-hand side of (3-31) only by bringing in a multiple of the delta function. We cannot get any differentiations of the delta function coming in, because the extra terms come from Poisson brackets of $\phi(q)$ with undifferentiated momentum variables. So that the only effect on the right-hand side of adding the term ϕ to $K_\perp$ can be adding on a multiple of the delta function. But the right-hand side has to be antisymmetrical between x and x', because the left-hand side is obviously antisymmetrical between x and x'. That prevents us from just adding a multiple of the delta function to the right-hand side of (3-31), so that it is *not* altered at all. Thus if the original $K_\perp$ satisfies the Poisson bracket relation (3-31), then the new one will also satisfy it.

There is a further factor which has to be taken into account to complete the proof. ϕ may also involve Γ $= \sqrt{-\det g_{rs}}$. One finds that $[w_\perp, \Gamma']$ involves $\delta(x - x')$ undifferentiated (one just has to work this out) and thus we can bring Γ into ϕ without disturbing the argument. In fact, we have to bring in Γ in order to preserve the validity of (3-22), which requires that $K_\perp^*$ and $K_\perp$ shall be scalar densities. We must thus bring in a suitable power of Γ to make ϕ a scalar density.

This is the method which is usually used in practice for bringing in interaction between fields without disturbing the relativistic character of the theory. For various simple fields the conditions turn out to be satisfied. We have the necessary bit of luck, and we can bring in interaction between fields of the simple character described and the conditions for the quantum theory to be relativistic are preserved.

There are some examples for which we *don't* have the necessary luck and we just cannot arrange the factors in $K_\perp$ to get (3-31) holding with the coefficients on the left, and then we do not know how to quantize the theory with states on curved surfaces. But actually, we are trying to do rather more than is necessary when we try to set up our quantum theory with states on curved surfaces. For the purposes of getting a theory in agreement with special relativity, it would be quite sufficient to have our states defined only on flat surfaces. That will involve some conditions on $K_\perp$ which are less stringent than those which I have formulated here. And it may be that we can satisfy these less stringent conditions without being able to satisfy those which I have formulated here.

An example for that is provided by the Born–Infeld electrodynamics, which is a modification of the Maxwell electrodynamics based on a different action integral, an action integral which is in agreement with the Maxwell one for weak fields, but differs from it for strong fields. This Born–Infeld electrodynamics leads to a classical $K_\perp$ which involves square roots. It is of such a nature that it doesn't seem possible to fulfill the conditions which are necessary for building up a relativistic quantum theory on curved surfaces. However, it does seem to be possible to build up a relativistic quantum theory on flat surfaces, for which the conditions are less stringent.

Lecture No. 4

QUANTIZATION ON FLAT SURFACES

We have been working with states on general space-like curved surfaces in space-time. I will just summarize the results that we obtained concerning the conditions for a quantum field theory, formulated in terms of these states, to be relativistic. We introduce variables to describe the surface, consisting of the four coordinates y^A of each point $x^r = (x^1, x^2, x^3)$ on the surface. The x's form a curvilinear system of coordinates on the surface. Then the y's are treated as dynamical coordinates and there are momenta conjugate to them, $w_A(x)$, again functions of the x's. And then we get a number of primary first-class constraints appearing in the Hamiltonian formalism, of the nature

$$w_A + K_A \approx 0. \qquad (3\text{-}10)$$

The K's are independent of the w's, but may be functions of any of the other Hamiltonian variables. The K's will involve the physical fields which are present. We analyze these constraints by resolving them into components tangential to the surface and normal to the surface. The tangential components are

$$w_r + K_r \approx 0, \qquad (4\text{-}1)$$

and the normal component is

$$w_\perp + K_\perp \approx 0. \tag{4-2}$$

With this analysis, we find that the K_r can be worked out just from geometrical considerations. The K_r should be looked upon as something rather trivial, associated with transformations in which the coordinates of the surface are varied, but the surface itself doesn't move. The first-class constraints (4-2) are associated with the motion of the surface normal to itself and are the important ones physically.

Certain Poisson bracket relations (3-21), (3-22), and (3-23) have to be fulfilled for consistency. Some of the Poisson bracket relations involve merely the K_r, and they are automatically satisfied when the K_r are chosen in accordance with the geometrical requirements. Some of the consistency conditions are linear in $K_\perp$ and they are automatically satisfied provided we choose $K_\perp$ to be a scalar density. Then finally we have the consistency conditions which are quadratic in the $K_\perp$ and those are the important ones, the ones which cannot be satisfied by trivial arguments.

These important consistency conditions can be satisfied in the classical theory if we work from a Lorentz-invariant action principle and calculate the $K_\perp$ by following the standard rules of passing from the action principle to the Hamiltonian. The problem of getting a relativistic quantum theory then reduces to the problem of suitably choosing the non-commuting factors which occur in the quantum $K_\perp$ in such a way that the quantum consistency conditions are fulfilled, which means that the commutator of two of the quantities (4-2) at two points in space x^1, x^2, x^3 has to be a linear combination of the constraints with coefficients occurring on the left. These quantum consistency conditions will usually be quite difficult to satisfy. It turns out that one can satisfy them with certain simple examples, but with more complicated examples it doesn't seem to be possible to satisfy them. That leads to the conclusion that one cannot set up a quantum theory for these more general fields with the states defined on general curved surfaces.

I might mention that the quantities K have a simple physical meaning. K_r can be interpreted as the momentum density, $K_\perp$ as the energy density; so the momentum density, expressed in terms of Hamiltonian variables, is something which is always easy to work out just from the geometrical nature of the problem and the energy density is the important quantity which one has to choose correctly (satisfying certain commutation relations) in order to satisfy the requirements of relativity.

If we cannot set up a quantum theory with states on general curved surfaces, it might still be possible to set it up with states defined only on flat surfaces.

We can get the corresponding classical theory simply by imposing conditions which make our previous curved surface into a flat surface. The conditions will be the following: The surface is specified by $y_A(x)$; in order to make the surface flat, we require that these functions shall be in the form

$$y_A(x) = a_A + b_{Ar}x^r, \qquad (4\text{-}3)$$

where the a's and b's are independent of the x's. This will result in the surface being flat, and in the system of coordinates x^r being rectilinear. At present we are not imposing the conditions that the x^r coordinate system shall be orthogonal: I shall bring that in a little later.

We are thus working with general, oblique, rectilinear axes x^r.

We now have our surface fixed by quantities a_Λ, $b_{\Lambda r}$ and these quantities will appear as the dynamical variables needed to fix the surface. We have far fewer of them than previously. In fact, we have only $4 + 12 = 16$ variables here. We have these 16 dynamical coordinates to fix the surface instead of the previous $y_\Lambda(x)$, which meant $4 \cdot \infty^3$ dynamical coordinates.

When we restrict the surface in this way, we may look upon the restriction as bringing a number of constraints into our Hamiltonian formalism, constraints which express the $4 \cdot \infty^3 y$ coordinates in terms of 16 coordinates. These constraints will be second-class. Their presence means a reduction in the number of effective degrees of freedom for the surface from $4 \cdot \infty^3$ to 16, a very big reduction!

In a previous lecture I gave the general technique for dealing with second-class constraints. The reduction in the number of effective degrees of freedom leads to a new definition of Poisson brackets. This general technique is not needed in our present case, where conditions are sufficiently simple for one to be able to use a more direct method. In fact, we can work out directly what effective momentum variables remain in the theory when we have reduced the number of effective degrees of freedom for the surface.

With our dynamical coordinates restricted in this way, we have of course the velocities restricted by the equation

$$\dot{y}_\Lambda = \dot{a}_\Lambda + b_{\Lambda r} x^r. \tag{4-4}$$

The dot refers to differentiation with respect to some parameter τ. As τ varies, this flat surface varies, moving parallel to itself and also changing its direction. The surface thus moves with a four-fold freedom, and this motion is expressed by our taking a_A, b_{Ar} to be functions of the parameter τ.

The total Hamiltonian is now

$$H_T = \int \dot{y}^A (w_A + K_A)\, d^3x$$

$$= \dot{a}^A \int (w_A + K_A)\, d^3x + \dot{b}_r^A \int x^r (w_A + K_A)\, d^3x. \quad (4\text{-}5)$$

(I have taken the quantities $\dot{a}^A$, $\dot{b}^A$. outside the integral signs, because they are independent of the x variables.) (4-5) involves the w variables only through the combinations $\int w_A\, d^3x$ and $\int x^r w_A\, d^3x$. We have here 16 combinations of the w's, which will be the new momentum variables conjugate to the 16 variables a, b which are now needed to describe the surface.

We can again express H_T in terms of the normal and tangential components of these quantities:

$$H_T = \dot{a}^A l_A \int (w_\perp + K_\perp)\, d^3x + \dot{a}^A b_{Ar} \int (w^r + K^r)\, d^3x$$

$$+ \dot{b}_r^A l_A \int x^r (w_\perp + K_\perp)\, d^3x + \dot{b}_{Ar} b_s^A \int x^r (w^s + K^s)\, d^3x.$$

$$(4\text{-}6)$$

Let us now bring in the condition that the x^r coordinate system is orthogonal. That means

$$b_{Ar} b_s^A = g_{rs} = -\delta_{rs}. \quad (4\text{-}7)$$

Differentiating (4-7) with respect to τ, we get

$$\dot{b}_{Ar} b_s^A + b_{As} \dot{b}_r^A = 0 \quad (4\text{-}8)$$

(I have been raising the Λ suffixes quite freely because the Λ coordinate system is just the coordinate system of special relativity.) This equation tells us that $b_{\Lambda r} b_s^\Lambda$ is antisymmetric between r and s. So the last term in (4-6) is equal to

$$\tfrac{1}{2} b_{\Lambda r} b_s^\Lambda \int \{x^r(w^s + K^s) - x^s(w^r + K^r)\}\, d^3x.$$

Now you see that we don't have so many linear combinations of the w's occurring in the H_T as before. The only linear combinations of the w's which survive are the following ones:

$$P_\perp \equiv \int w_\perp\, d^3x, \tag{4-9}$$

$$P_r \equiv \int w_r\, d^3x, \tag{4-10}$$

and also
$$M_{r\perp} \equiv \int x^r w_\perp\, d^3x, \tag{4-11}$$

and
$$M_{rs} \equiv \int (x_r w_s - x_s w_r)\, d^3x. \tag{4-12}$$

(We can raise and lower the suffixes r quite freely now because they refer to rectilinear orthogonal axes.) These are the momentum variables which are conjugate to the variables needed to fix the surface when the surface is restricted to be a flat one referred to rectilinear orthogonal coordinates.

The whole set of momentum variables included in (4-9), (4-10), (4-11), and (4-12) can be written as P_μ and $M_{\mu\nu} = -M_{\nu\mu}$, where the suffixes μ and ν take on 4 values, a value 0 associated with the normal component, and 1, 2, 3 associated with the three x's. μ, ν are small suffixes referring to the x coordinate system, to distinguish them from the capital suffixes Λ referring to the fixed y coordinate system.

So now our momentum variables are reduced to just 10 in number, and associated with these 10 momentum variables we have 10 primary first-class constraints, which we may write

$$P_\mu + p_\mu \approx 0, \qquad (4\text{-}13)$$

$$M_{\mu\nu} + m_{\mu\nu} \approx 0, \qquad (4\text{-}14)$$

where
$$p_\perp \equiv \int K_\perp \, d^3x, \qquad (4\text{-}15)$$

$$p_r \equiv \int K_r \, d^3x, \qquad (4\text{-}16)$$

$$m_{r\perp} \equiv \int x_r K_\perp \, d^3x, \qquad (4\text{-}17)$$

and
$$m_{rs} \equiv \int (x_r K_s - x_s K_r) \, d^3x. \qquad (4\text{-}18)$$

We have now 10 primary first-class constraints associated with a motion of the flat surface. In Lecture (3) I said that we would need 4 primary first-class constraints (3-10) to allow for the general motion of a flat surface. We see now that 4 is not really adequate. The 4 has to be increased to 10, because 4 elementary motions of the surface normal to itself and changing its direction would not form a group; in order to have these elementary motions forming a group, we have to extend the 4 to 10, the extra 6 members of the group including the translations and rotations of the surface, which motions affect merely the system of coordinates in the surface without affecting the surface as a whole. In this way we are brought to a Hamiltonian theory involving 10 primary first-class constraints.

We have now to discuss the consistency conditions, the conditions in terms of Poisson bracket relations which are necessary for all the constraints to be first-class. Let us first discuss the Poisson bracket relations between the momentum variables P_μ, $M_{\mu\nu}$. We are given these momentum variables in terms of the w variables (4-9) to (4-12), and we know the Poisson bracket relations (3-17), (3-18), and (3-19) between the w variables, so we can calculate the Poisson bracket relations between the P and M variables. It is not really necessary to go through all this work to determine the Poisson bracket relations between the P and M variables. It is sufficient to realize that these variables just correspond to the operators of translation and rotation in four-dimensional flat space-time, and thus their Poisson bracket relations must just correspond to the commutation relations between the operators of translation and rotation. In either way we get the following Poisson bracket relations:

$$[P_\mu, P_\nu] = 0 \qquad (4\text{-}19)$$

which expresses that the various translations commute;

$$[P_\mu, M_{\rho\sigma}] = g_{\mu\rho}P_\sigma - g_{\mu\sigma}P_\rho; \qquad (4\text{-}20)$$

and $[M_{\mu\nu}, M_{\rho\sigma}] = -g_{\mu\rho}M_{\nu\sigma} + g_{\nu\rho}M_{\mu\sigma} + g_{\mu\sigma}M_{\nu\rho} - g_{\nu\sigma}M_{\mu\rho}.$ $(4\text{-}21)$

Let us now consider the requirements for the equations (4-13) and (4-14) to be first-class. The Poisson bracket of any two of them must be something which vanishes weakly and must therefore be a linear combination of them. So we are led to these Poisson bracket relations:

$$[P_\mu + p_\mu, P_\nu + p_\nu] = 0, \qquad (4\text{-}22)$$

$$[P_\mu + p_\mu, M_{\mu\sigma} + m_{\mu\sigma}] = g_{\mu\rho}(P_\sigma + p_\sigma) \\ - g_{\mu\sigma}(P_\rho + p_\rho), \quad (4\text{-}23)$$

and $[M_{\mu\nu} + m_{\mu\nu}, M_{\rho\sigma} + m_{\rho\sigma}] = -g_{\mu\rho}(M_{\nu\sigma} + m_{\nu\sigma})$
$+ g_{\nu\rho}(M_{\mu\sigma} + m_{\mu\sigma}) + g_{\mu\sigma}(M_{\nu\rho} + m_{\nu\rho}) - g_{\nu\sigma}(M_{\mu\rho} + m_{\mu\rho}).$
$$(4\text{-}24)$$

The argument for getting these relations is that, on the right-hand sides we had to put something which is weakly equal to zero in each case, and we know the terms on the right-hand sides which involve the momentum variables P, M because these terms come only from the Poisson brackets of momenta with momenta and so are given by (4-19), (4-20), and (4-21). (I have already used the same argument in the curvilinear case for (3-21), (3-22), and (3-23), so there is no need to go into detail here. For example, see how (4-23) comes about. The terms involving P are just the same as in (4-20). They come from the Poisson bracket of P and M. The remaining terms are filled in in order to make the total expression weakly equal to zero.) (4-22), (4-23), and (4-24) are the requirements for consistency.

We can make a further simplification, which we could not do in the case of curvilinear coordinates, in this way: Let us suppose that our basic field quantities are chosen to refer only to the x coordinate system. They are field quantities at specific points x in the surface, and we can choose them so as to be quite independent of the y coordinate system. Then the quantities K_1, K_r will be quite independent of the y coordinate system, and that means that they will have zero Poisson brackets with the variables P, M. We then have a zero Poisson bracket between each of the variables p, m and each of the P, M.

This condition follows with the natural choice of dynamical variables to describe the physical fields which are present. We cannot do the corresponding simplification when we are working with the curved surfaces, because the g_{rs} variables that fix the metric will enter into the quantities $K_\perp$, K_r. The result is that we cannot set them up in a form which does not refer at all to the y coordinate system, because the y coordinates enter into the g_{rs} variables. However, with the flat surfaces, we can make this simplification, and that results in equations (4-22), (4-23), and (4-24) simplifying to

$$[p_\mu, p_\nu] = 0; \qquad (4\text{-}25)$$

$$[p_\mu, m_{\rho\sigma}] = g_{\mu\rho}p_\sigma - g_{\mu\sigma}p_\rho; \qquad (4\text{-}26)$$

and

$$[m_{\mu\nu}, m_{\rho\sigma}] = -g_{\mu\rho}m_{\nu\sigma} + g_{\nu\rho}m_{\mu\sigma} + g_{\mu\sigma}m_{\nu\rho} - g_{\nu\sigma}m_{\mu\rho}.$$
$$(4\text{-}27)$$

P and M have disappeared from these equations, so the consistency conditions now involve only the field variables, and not the variables, which are introduced for describing the surface. In fact, these conditions merely say that the p, m shall satisfy Poisson bracket relations corresponding to the operators of translation and rotation in flat space-time. The problem of setting up a relativistic field theory now reduces to finding the quantities p, m to satisfy the Poisson bracket relations (4-25), (4-26), and (4-27).

These quantities, remember, are defined in terms of $K_\perp$ and K_r, the energy density and the momentum

[76]

density. The expression for the momentum density is just the same as in curvilinear coordinates. It is determined by geometrical arguments only. Our problem reduces to finding the energy density $K_\perp$ leading to p's and m's such that the Poisson bracket relations (4-25), 4-26), and (4-27) are fulfilled.

If we work from a Lorentz-invariant action integral and deduce $K_\perp$ from it by standard Hamiltonian methods, $K_\perp$ will automatically satisfy these requirements in the classical theory. The problem of getting a relativistic quantum theory then reduces the problem of suitably choosing the order of factors which occur in $K_\perp$ so as to satisfy the equations (4-25), (4-26), and (4-27) also in the quantum theory, where the Poisson bracket becomes a commutator and the p, m involve non-commuting quantities.

Let us look at (4-25), (4-26), and (4-27) and substitute for p and m their values in terms of K's. Then you see that some of these conditions will be independent of $K_\perp$. These are automatically satisfied when we choose K_r properly, in accordance with the geometrical requirements. Some of the conditions are linear in $K_\perp$. These will be satisfied by taking $K_\perp$ to be any three-dimensional scalar density in the space of the x's. So that there is no problem in satisfying the conditions which are linear in $K_\perp$. The awkward ones to satisfy are the ones which are quadratic in $K_\perp$. They are the following:

$$\left[\int x_r K_\perp \, d^3x, \int K_\perp' \, d^3x' \right] = \int K_r \, d^3x. \quad (4\text{-}28)$$

(This equation comes from (4-26) where we put $\mu = \perp$, $\rho = r$, and $\sigma = \perp$.)

$$\left[\int x_r K_\perp \, d^3x, \int x'_s K'_\perp \, d^3x' \right] = - \int (x_r K_s - x_r K_r) \, d^3x$$

$$(4\text{-}29)$$

(from (4-27) where we take $\nu = \perp$ and $\sigma = \perp$). So the problem of getting a relativistic quantum field theory now reduces to the problem of finding an energy density $K_\perp$ which satisfies the conditions (4-28) and (4-29) when we take into account non-commutation of the factors.

We can analyze these conditions a little more when we take into account that the Poisson bracket connecting $K_\perp$ at one point and $K'_\perp$ at another point will be a sum of terms involving delta functions and derivatives of delta functions:

$$[K_\perp, K'_\perp] = a\delta + 2b_r\delta_{,r} + c_{rs}\delta_{,rs} + \dots . \quad (4\text{-}30)$$

(This delta is the three-dimensional delta function involving the three coordinates x and the three coordinates x' of the first and second points.) Here $a = a(x)$, $b = b(x)$, $c = c(x), \dots$ One could have the coefficients involving also x', but then one could replace them by coefficients involving x only at the expense of making some changes in the earlier coefficients in the series. There is no fundamental dissymmetry between x and x', only a dissymmetry in regard to the way the equation is written.

(4-30) is the general relationship connecting the energy density at two points. Now for many examples, including all the more usual fields, derivatives of the delta function higher than the second do not occur. Let us examine this case further.

Assume derivatives higher that the second do not occur. That means that the series (4-30) stops at the third term. In this special case we can get quite a bit of information about the coefficients a, b, c by making use of the condition that the Poisson bracket (4-30) is anti-symmetrical between the two points x and x'. Interchanging x and x' in (4-30), we get

$$[K'_\perp, K_\perp] = a'\delta - 2b'_r\delta_{,r} + c'_{rs}\delta_{,rs}$$
$$= a'\delta - 2(b'_r\delta)_{,r} + (c'_{rs}\delta)_{,rs}$$

(since $\partial b_r(x')/\partial x^r = 0$, etc.)

$$= a\delta - 2(b_r\delta)_{,r} + (c_{rs}\delta)_{,rs}$$
$$= (a - 2b_{r,r} + c_{rs,rs})\delta + (-2b_r + 2c_{rs,r})\delta_{,r}$$
$$+ c_{rs}\delta_{,rs}. \qquad (4\text{-}31)$$

The expression (4-31) must equal minus the expression (4-30) identically. In order that the coefficients of $\delta_{,rs}$ shall agree we must have

$$c_{rs} = 0. \qquad (4\text{-}32)$$

This then makes the coefficients of $\delta_{,r}$ agree. Finally, in order that the coefficients of δ shall agree, we must have

$$a = b_{r,r}. \qquad (4\text{-}33)$$

This gives us the equation

$$[K_\perp, K'_\perp] = 2b_r\delta_{,r} + b_{r,r}\delta. \qquad (4\text{-}34)$$

Let us now substitute in (4-28) and (4-29). They become:

$$\int K_r \, d^3x = \iint x_r(2b_s\delta_{,s} + b_{s,s}\delta) \, d^3x \, d^3x'$$
$$= \int x_r b_{s,s} \, d^3x$$
$$= \int b_r \, d^3x. \qquad (4\text{-}35)$$

(Note that $x_{r,s} = \partial x_r/\partial x^s = -\delta_{rs}$.)

[79]

$$- \int (x_r K_s - x_s K_r) \, d^3x = \iint x_r x_s' (2b_t \delta_{,t} + b_{t,t} \delta) \, d^3x \, d^3x'$$

$$= \int (-2x_r b_s + x_r x_s b_{t,t}) \, d^3x$$

$$= \int (-x_r b_s + x_s b_r) \, d^3x. \qquad (4\text{-}36)$$

This is what our consistency conditions reduce to, and we see that they are satisfied by taking $b_r = K_r$. This is not quite the most general solution; more generally we could have

$$b_r = K_r + \theta_{rs,s} \qquad (4\text{-}37)$$

for any quantity θ_{rs} satisfying the condition that

$$\int (\theta_{rs} - \theta_{sr}) \, d^3x = 0. \qquad (4\text{-}38)$$

Thus θ can have any symmetrical part and its anti-symmetrical part must be a divergence.

That gives the general requirement for a field theory to be relativistic. We have to find the energy density $K_\perp$ satisfying the Poisson bracket relation (4-34) where b_r is connected with the momentum density by (4-37). If we work out the energy density from a Lorentz-invariant action then this condition will certainly be satisfied in the classical theory. It might not be satisfied in the quantum theory because the order of the factors might be wrong. It is only when one can choose the order of the factors in the energy density so as to make (4-34), (4-37) hold accurately that we have a relativistic quantum theory. The conditions which we have here for a quantum theory to be relativistic are less stringent than the ones which we obtained when we had states defined on general curved surfaces.

I would like to illustrate that by taking the example of Born–Infeld electrodynamics. This is an electrodynamics which is in agreement with Maxwell electrodynamics for weak fields but differs from it for strong fields. (We now refer the electromagnetic field quantities to some absolute unit defined in terms of the charge of the electron and classical radius of the electron, so that we can talk of strong fields and weak fields.) The general equations of the Born–Infeld electrodynamics follow from the action principle:

$$I = \int \sqrt{- \det (g_{\mu\nu} + F_{\mu\nu})} \, d^4x. \qquad (4\text{-}39)$$

We may use curvilinear coordinates at this stage. $g_{\mu\nu}$ gives the metric referred to these curvilinear coordinates and $F_{\mu\nu}$ gives the electromagnetic field referred to the absolute unit.

We can pass from this action integral to a Hamiltonian by using the general procedure. The result is to give us a Hamiltonian in which we have, in addition to the variables needed to describe the surface, the dynamical coordinates A_r, $r = 1, 2, 3$. A_0 turns out to be an unimportant variable just like in the Maxwell field. The conjugate momenta D^r to the A_r are the components of the electric induction, and satisfy the Poisson bracket relations

$$[A_r, D'^s] = g_r^s \delta(x - x'). \qquad (4\text{-}40)$$

It turns out that in the Hamiltonian we only have A occurring through its curl, namely through the field quantities:

$$B^r = \tfrac{1}{2}\varepsilon^{rst}F_{st} = \varepsilon^{rst}A_{t,s}. \qquad (4\text{-}41)$$

$\varepsilon^{rst} = 1$ when $(rst) = (1, 2, 3)$ and is anti-symmetrical between the suffixes. The commutation relation between B and D is

$$[B^r, D'^s] = \varepsilon^{rst}\delta_{,t}(x - x'). \qquad (4\text{-}42)$$

The momentum density now has the value

$$K_r = F_{rs}D^s. \qquad (4\text{-}43)$$

This is just the same as in the Maxwell theory. It is in agreement with the general principle that the momentum density depends only on geometrical arguments, i.e. on the geometrical character of the fields we are using, and the action principle doesn't matter.

The energy density now has the value

$$K_\perp = \{\varGamma^2 - \gamma_{rs}(D^rD^s + B^rB^s) - \gamma^{rs}F_{rt}F_{su}D^tD^u\}^{1/2} \qquad (4\text{-}44)$$

Here γ_{rs} is the metric in the three-dimensional surface and

$$-\varGamma^2 = \det \gamma_{rs}. \qquad (4\text{-}45)$$

If we work with curved surfaces we require $K_\perp$ to satisfy the Poisson bracket relation (3-31). In the classical theory it must do so because it is deduced from a Lorentz-invariant action integral. But we cannot get it to satisfy the required commutation relationship in the quantum theory. The expression for $K_\perp$ has a square root occurring in it, which makes it very awkward to work with. It seems to be quite hopeless to try to get the commutation relations correctly fulfilled with the coefficients γ^{rs} occurring on the left. So it does not seem to be possible to get a Born–Infeld quantum electro-

[82]

dynamics with the state defined on general curved surfaces.

Let us, however, go over to flat surfaces. For that purpose, we need to work out the Poisson bracket relationship (4-34). Now we know that conditions are all right in the classical theory. Classically we must therefore have the Poisson bracket relationship:

$$[K_\perp, K'_\perp] = 2K_r\delta_{,r} + K_{r,r}\delta. \qquad (4\text{-}46)$$

We can see without going into detailed calculations that this must hold also in the quantum theory, because K_r is built up entirely from the quantities D^s and B^t. When we work things out in the quantum theory, we shall have the D's and the B's occurring in a certain order, but the D's and B's all commute with each other when we take them at the same point. We see that from (4-42). If we put the $x' = x$ we get

$$[B^r, D^s] = \varepsilon^{rst}\delta_{,t}(0) = 0 \qquad (4\text{-}47)$$

(the derivative of the delta function with the argument 0 is to be taken as zero). Thus we are not bothered by the non-commutation of the D's and the B's that occur in K_r. We must therefore get the classical expression, so that the consistency conditions *are* fulfilled.

So for the Born–Infeld electrodynamics, the consistency conditions for the quantum theory on flat surfaces are fulfilled, while they are not fulfilled on curved surfaces. Physically that means that we can set up the basic equations for a quantum theory of the Born–Infeld electrodynamics agreeing with special relativity, but we should have difficulties if we wanted to have this quantum theory agreeing with general relativity.

That completes the discussion of the consistency

[83]

requirements for the quantum theory to be relativistic. However, even if we have satisfied these consistency requirements, we have not yet disposed of all the difficulties. There are some quite formidable difficulties still lying ahead of us. If we were dealing with a system involving only a finite number of degrees of freedom, then we should have disposed of all of the difficulties, and it would be a straightforward matter to solve the differential equations on ψ. But with field theory, we have an infinite number of degrees of freedom, and this infinity may lead to trouble. It usually does lead to trouble.

We have to solve equations in which the unknown, the wave function ψ, involves an infinite number of variables. The usual method that people have for solving this kind of equation is to use perturbation methods in which the wave function is expanded in powers of some small parameter, and one tries to get a solution step by step. But one usually runs into the difficulty that after a certain stage the equations lead to divergent integrals.

People have done a great deal of work on this problem. They have found methods for handling these divergent integrals which seem to be tolerable to physicists even though they cannot be justified mathematically, and they have built up the renormalization technique, which allows one to disregard the infinities in the case of certain kinds of field theory.

So, even when we have formally satisfied the consistency requirements, we still have the difficulty that we may not know how to get solutions of the wave equation satisfying the required supplementary conditions. If we can get such solutions, there remains the further problem of introducing scalar products for these

[84]

solutions, which means considering these solutions as the vectors in a Hilbert space. It is necessary to introduce these scalar products before we can get a physical interpretation for our wave function in terms of the standard rules for the physical interpretation of quantum mechanics. It is necessary that we should have scalar products for the wave functions which satisfy the supplementary conditions, but we do not need to worry about scalar products for general wave functions which do not satisfy the supplementary conditions. There may be no way of defining scalar products for these general wave functions, but that would not matter at all. The physical interpretation for quantum mechanics requires that scalar products exist only for wave functions satisfying all the supplementary conditions.

You see that there are quite formidable difficulties in getting the Hamiltonian theory to work, in connection with quantum mechanics. So far as concerns classical mechanics, the method seems to be fairly complete and we know exactly what the situation is; but for quantum mechanics we have only really started on the problem. There are the difficulties of finding solutions even when the supplementary conditions are formally consistent, and possibly also the difficulty of introducing scalar products of the solutions.

The difficulties are quite serious, and they have led some physicists to challenge the whole Hamiltonian method. A good many physicists are now working on the problem of trying to set up a quantum field theory independently of any Hamiltonian. Their general method is to introduce quantities which are of physical importance, then to bring in accepted general principles in order to impose conditions on these quantities; and their hope is that ultimately they will get enough conditions imposed on these quantities of physical importance to be able to calculate them. They are still very far from achieving that end, and my own belief is that it will not be possible to dispense entirely with the Hamiltonian method. The Hamiltonian method dominates mechanics from the classical point of view. It may be that our method of passing from classical mechanics to quantum mechanics is not yet correct. I still think that in any future quantum theory there will have to be something corresponding to Hamiltonian theory, even if it is not in the same form as at present.

I have given the treatment of the Hamiltonian method as far as it has yet been developed. It is quite a general and powerful method which can be adapted to a variety of problems. It can be adapted to problems where singularities (point or surface) occur in the field. The general idea governing this development of the Hamiltonian theory is to find an action I which involves certain parameters q, such that when we vary the q's, δI is linear in the δq's. It is indispensable that we should have δI linear in the δq's in order that we may apply the treatment described in these lectures.

The way to bring about linearity when we have singularities is to work in terms of curvilinear coordinates, and not to vary any equations which determine the position of a singular point or a singular surface. For example, if we are dealing with a singular surface specified by an equation $f(x) = 0$, then we must have a variation principle in which $f(x)$ is not varied. If we allow $f(x)$ to vary, if we treat f itself as providing some of the q's, then we do not have δI linear in the δq's. But we can keep $f(x)$ fixed with respect to some curvilinear coordinate system x and we can vary the surface by varying the curvilinear coordinate system without varying the function f. Then the general method which I have discussed here works very well in the classical theory. When we go over to quantization we have the difficulties arising which I have discussed.